Biological Education
for
Community Development

Commission for Biological Education of the International Union of Biological Sciences

The Commission is appointed by the International Union of Biological Sciences. It consists currently of people concerned with biological education from twenty-one countries covering all continents. Members serve for periods of three years so that the opportunity is given for people from other countries to take the place of those retiring. Its interests cover school education, tertiary education including teacher education, and non-formal education.

A Newsletter is published from Yarmouk University, Irbid, Jordan, and edited by Professor A. Badran.

The present members of the Commission are as follows:

Executive: Professor P. J. Kelly, England (Chairman), Professor A. Badran, Jordan (Vice Chairman), Professor R. Kille, Scotland (Ex-Chairman), Professor H. Camefort, France (Secretary/Treasurer), Professor A. Grebecki, Poland, Professor A. Rao, Singapore.

Other members: Dr. M. Atchia, Mauritius, Professor V. Basnayake, Sri Lanka, Professor J. M. Gutierrez-Vasquez, Mexico, Professor S. L. Gueye, Gabon, Professor F. Gornall, Canada, Professor D. Hernandez, Phillippines, Professor M. Chihara, Japan, Mr. W. King, Barbados, Professor M. Krasilchik, Brazil, Professor W. M. Laetsch, USA, Professor D. Lockard, USA, Professor W. Mayer, USA, Professor R. Meyer, Australia, Professor C. Quiroz, Peru, Mr. E. B. Rugumayo, Uganda, Dr. Saber-Selim, Egypt, Professor G. Schaefer, West Germany.

Acknowledgements

The Division of Higher Education and Research, and of Science, Technical and Vocational Education of UNESCO have both encouraged and generously supported the work of the IUBS Commission for Biological Education. At the London meeting we were also supported by the British Council.

This publication was additionally supported by a special grant from the Division of Higher Education and Research and of Science, Technical and Vocational Education of UNESCO.

Biological Education for Community Development

Edited by

P. J. Kelly and G. Schaefer

on behalf of the

Commission for Biological Education
of the International Union of
Biological Sciences (IUBS)

and in cooperation with the

Institute for Science Education (IPN)
at the University of Kiel, West Germany

Taylor & Francis Ltd
London
1980

First published 1980 by Taylor & Francis Ltd, 4 John Street,
London WC1N 2ET.

ISBN 0 85066 214 1

Printed and bound in Great Britain by Taylor & Francis (Printers) Ltd.
Rankine Road, Basingstoke, Hampshire RG24 0PR

Contents

PREFACE

Many of the problems facing countries throughout the world are essentially biological in nature. Nutrition, health, agriculture, pollution, population growth, aggression, and the conservation of natural resources are examples of aspects of the life of a community which require an understanding of biology, if they are to be dealt with realistically. With advances in biological science, such as those related to biological technology and genetic engineering, have come further significant influences on peoples' lives.

Such considerations have led the Commission for Biological Education of the International Union of Biological Sciences (CBE-IUBS) to concentrate its work over the last three years on the problem of relating biological education to community development. In 1978 an international workshop was organized at Chelsea College in the University of London, England. At this the participants attempted to define what was meant by the notion of "biological education for community development". This workshop was followed by a further one at the Institute for Science Education (IPN) at the University of Kiel, West Germany, in 1979, with the establishment of three projects concerned with developing methods and materials for the three levels of school education, tertiary education, and non-formal education.

Brief accounts of these activities have been outlined in the Commission's Newsletters No. 2 and 4.[1] In this book we present a collection of papers and comments from the two workshops. It is a unique document in that it contains information and ideas on the subject from a truly international perspective.

Our discussions were focussed initially on a working paper "Biological Education for Community Development" which now is the first article of section I of this book. Section I contains also a collection of participants' reflections on the ideas it contains. The section presents a variety of interpretations but it will be clear to the reader that there are at least two common threads in the arguments: first that in education biological topics should be portrayed, whenever relevant, in their social context; and second that, besides an experience of scientific methods, students of biology should be introduced to the nature of problem-solving and decision-making relevant to dealing with social issues.

Section II includes accounts of the way biological education is being related to community needs in several countries. From this we can see that whilst much is being undertaken throughout the world, there is still much to be done, and this applies to developing countries as well as to those which are labelled as "developed".

In section III the papers deal with a number of particular issues

---

1) Editor: Professor D.A.M. BADRAN. Published at Yarmouk University, Irbid, Jordan, Jan. 1979 and March 1980.

concerning biological education for community development. They are witness of the complexity of the problem we are concerned with and are examples of the large variety of things to be done.

Section IV, finally, brings some more data and tables of current projects which may round up the picture of this book.

The views expressed in the articles are, of course, those of the authors. As editors we have attempted only to relate them to each other and to add clarity. The book is published on behalf of the IUBS Commission for Biological Education and IPN Kiel and it is the hope of both that it will provide ideas and stimulate action on what is probably the most significant question facing biology teachers today: How can biological education serve the combined needs of the individual, of society, and of science within it?

Southampton and Kiel, September 1, 1980

Peter J. Kelly Gerhard Schaefer

List of Contributors

Atchia, M.	Dr., Senior Lecturer and Science Section Co-ordinator, Mauritius Institute of Education, Reduit, Mauritius
Bergmann, H.,	Dr., GTZ, Forschungsstelle für Internationale Agrarentwicklung (Research Institute for International Rural Development) of Heidelberg
Blum, A.,	Dr., Director, Department of Agriculture, Hebrew University of Jerusalem, Rehovot/Israel
Dowdeswell, W.H.,	Professor of Education, University of Bath, School of Education, Claverton Down, Bath BA2 7AY
Gornall, F.A.,	Professor of Science Education, University of British Columbia, Vancouver 8, Canada
Gutiérrez-Vasquez, J.M.,	Professor, Head of the Department of Educational Research, Mexico
Hernandez, Dolores,	Professor, Head of Science Education Center (UPSEC), University of the Philippines, Quezon City, Manila, Philippines
Imahori, K.,	Professor, Dept. of Biology, College of General Education, Osaka University, Osaka 560/Japan
Kattmann, U.,	Dr., Department of Biology Education, IPN (Institute for Science Education), University of Kiel, West Germany
Kelly, P. J.	Professor of Education, Head of Department of Education, University of Southampton, England
Kille, R. A.,	Professor of Zoology, Director of Biology Teaching Unit, Department of Zoology, University of Edinburgh, Scotland
King, W. K.,	Lecturer in Science Education, University of the West Indies, School of Education, Bridgetown, Barbados
Krasilchik, Myriam,	Professor, Universidade de Sao Paulo, Cidade Universitaria, Sao Paulo, Barzil
Quiroz, C.A.,	Professor, National Coordinator of NIDE, Coraceros 158, Lima 21, Peru
Rao, A.N.,	Professor of Botany, Department of Botany, University of Singapore, Singapore 10
Rugumayo, E.,	formerly Lecturer of Education, School of Education, University of Zambia, P.O. Box 2379, Lusaka, Zambia
Saber Selim, M.,	Dr., Director, Science Department, ALECSO, The Arab Educational, Cultural and Scientific Organization, Cairo, Egypt
Schaefer, G.	Professor, Head of Department of Biological Education at IPN (Institute for Science Education), University of Kiel, Kiel, West Germany
Subbarini, M.S.	Biology Supervisor, Ministry of Education, Kuwait
Sutton, C.	Senior Lecturer in Education, School of Education, The University of Leicester, Leicester, England
Tomley, D.	Lecturer in Education, School of Education, The University, Leicester/England
Vlijm, L.	Professor, Institute of Ecology, Vrije Universiteit Amsterdam, De Boelelaan 1087, Amsterdam, Netherlands. Chairman of ECBA (European Communities' Biologists Association)

Section I

What is Biological Education for Community Development?

WORKING DOCUMENT FOR A MEETING ON BIOLOGICAL EDUCATION FOR COMMUNITY DEVELOPMENT

by P. J. Kelly

The meeting was commissioned and financed by UNESCO and organized by the Commission for Biological Education of the International Union of Biological Sciences (IUBS) at Chelsea College, University of London, England, December 16-20, 1978.

The meeting had four general objectives.

1. To define the relationships between biological education and community needs.
2. To analyse the content, skills (especially those of decision-making), and attitudes involved in biological education related to community development.
3. To work out practical activities which can enable schools, teacher education institutions, and community agencies to achieve a closer and more beneficial relationship between biological studies and community development.
4. To define the particular roles that tertiary education and research in biology can play in community development and suggest practical activities to foster them.

Some Background Thoughts. What follows here is a personal perspective of some of the issues I see as important for our discussion of the relation between biological education and community development. Hopefully, it will, at the least, stimulate our thoughts and produce ideas.

Scientific biology and community biology.
Some fifteen years or so ago the emphasis in biological education at the school level was on the presentation of information about living things: the descriptive approach. In more recent times there has been a strong movement to change courses so that they depict biology more as an experimental science with pupils acquiring some of the skills of scientific investigation and, in part at least, obtaining biological knowledge through such investigations. This can be termed the scientific enquiry approach.

Now we are saying that biological education should be related to community needs and so we have to ask ourselves whether this is in any way different from these previous approaches.

I would like to suggest three ways in which I feel there is a difference. First, a community approach requires that those biological topics which have implications for the community have greater priority for inclusion in the school curriculum. Such topics would presumably include those related to health, environment (ecology), nutrition, population, genetics, behaviour and the biological technologies such as those of agriculture, fisheries and medicine.

Second, there would be a greater emphasis on human biology. Third, there would be a greater emphasis on what may be called issue studies rather than knowledge studies.[1] Examples of topics

[1] See page 30 (Science, Life and Education, page 5)

for knowledge studies are Mendelian inheritance and the food requirements of the body. They would be studied by the descriptive or scientific enquiry methods. Their equivalent issue studies are genetic counselling and planning the provision of a balanced diet from locally available food crops. Such studies would involve value judgments, problem-solving involving a range of possible answers, the skills of social decision-making and other considerations besides scientific judgments. They would require a more interdisciplinary approach incorporating aspects of, for example, economics and cultural traditions. Issue studies, in fact, involve similar skills and attitudes to those needed generally in dealing with the problems of community development.

It may well be important to remind ourselves, however, that whilst the community approach to biological education is clearly of great importance and its inclusion in school courses imperative, there is still an important place for scientific and descriptive approaches. The problem is to achieve the right balance between the three approaches.

Educational requirements for community development. Besides considering what aspects of biological education *we* feel are of value to a community we should also put ourselves in the place of people in a community and ask what *they* would require from education in order that they can participate in development. I suspect that an answer to such a question should have three dimensions: creating awareness, fostering personal development and developing the skills of social action.

A prime function of education should be to heighten a community's *awareness* of what biological understanding is necessary for adequate development and demonstrate the extent to which it can lead to improvement in the quality of life. The mere provision of information will not satisfy this need. It will have to be in a form that relates to immediate and local aspects of development and involve direct studies of local situations and/or case studies of other, but similar, localities which will demonstrate the potential value of development to the community.

The effectiveness of a person's contribution to community development depends on their personal commitment to it. Whilst this may be motivated by an altruistic desire to be of service to the community, it will no doubt, be stronger if it is seen as affecting the *personal development* of individuals. For example, if a person has an understanding of their own body and its requirements so that they have a sense of the potential and limitations of their physiology and behaviour and hence of being in control of their personal development, then they are more likely to commit themselves to community developments related, say, to health or nutrition. Invariably schemes linking biology to community development are formulated in terms of social biology. To my mind an adequate treatment of the biology of the individual person is equally important.

Given a commitment to community development a person will also need to acquire the skills and attitudes of community action. How can he (or she) define his role as a participant and realise the skills of cooperation, of decision-making and of evaluating the validity of information and attitudes he will encounter? Although real practical experience must obviously be the ultimate answer to such questions, it may also be argued that education aimed at helping people to be aware of the nature of such skills and attitudes and which provides opportunity for practising them in simulated or small-scale practical activities, is also necessary.

Schools and community education. Schools are specialized agencies of education. They are not, however, the only educational agencies in a community, nor are they necessarily the most influential in supporting community development. What a child learns from his parents or through his out-of-school-activities, for example, may have as much, if not more, influence on his feeding habits, his eventual attitudes to health and conservation programmes and his participation in community activities, as anything he does in school. Thus, in discussing education for community development we will need to consider pupils within their total community cultures, not just as participants in the education system.

Bearing this in mind we could consider devising projects which, although they have similar aims attempt to achieve them in different ways. I will outline five types of project that seem to me to be relevant.

a) Development of course modules of materials and activities which provide pupils with the basic knowledge and skills for biological community development. The modules should be sufficiently adaptable to enable teachers to fit them into their current school curricula and relate them to natural or local circumstances. The intention of this type of project would be primarily to prepare pupils for participation in community development when they leave school; an aim similar to that of much of the school curriculum.

b) Development of materials and ideas for activities related directly to the pupils' needs and their own roles in community development to be used primarily out-of-school although this would not preclude them from being linked to school work. Magazines and packages of activities suitable for youth groups are examples of such materials. Some thought would have to be given to preparing materials for youngsters of low literacy.

c) Development of materials and ideas for activities which are aimed primarily at the adult population e.g. parents, youth leaders, doctors, nurses and farmers. These would be aimed at educating the adults but provide them also with ideas and materials to enable them to undertake activities with the youngsters they are associated with.

d) Development of strategies for joint school-community activity. This could involve teachers (and possibly senior pupils, if only as observers) as members of local environmental, health or similar groups concerned with some aspect of community development. The teachers could then use this experience in their teaching and undertake activities with their pupils related to the community work. In return the school facilities could be utilised by the group.
e) Development of strategies for total community activity. For this the school is seen as only one of several community agencies which might be concerned with community development. A strategy would be based on a single topic e.g. nutrition education. There would need to be an initial research phase when an analysis is made of the perceptions and attitudes of different groups in the community and of the agencies (actual and potential) which were relevant to the topic. Following this a variety of educative activities could be developed in the community based on a number of agencies, some of which might be schools if they were particularly appropriate.

Tertiary education, research and community development.

Tertiary education may be divided into three categories -

1. degree level studies taking place in universities, polytechnics and colleges where, to varying extents, research is also pursued
2. non-degree vocational studies pursued in colleges e.g. technical, nursing and commercial colleges, where it is not customary for research to take place
3. continuing general studies pursued in some form of adult education system.

All these are relevant to our discussions and much of what was said earlier in this paper applies to them. However, those forms of tertiary education which come under 1. and 2. have particular values to which, I suggest, we should give special consideration. In our universities and colleges rests a considerable store-house of expertise of value to community development and from their students come, on the whole, the significant decision-makers of community development.

Taking this last point first we could consider what types of content, materials and teaching methods are required for courses (or modules of courses) in tertiary education which can both inform students and train them in the relevant decision-making skills for biological community development. Such courses could be aimed at the non-biologists - particularly those who may administer community development later - as well as students taking biological or allied studies and, of course, student biology teachers.

Using the expertise of universities and colleges (and one might add here research institutes, botanical gardens and similar establishments whose primary function is not usually educational) for community development may be looked at in a number of ways.

What are the most suitable methods of disseminating the relevant knowledge of universities and colleges to community development programmes? What systems of information storage and retrieval are required in colleges and universities in order that those working in community development can readily obtain the help they require?

To what extent can universities and colleges serve as agencies of public education in relation to biological community development? Should botanical gardens, farms, community health centres etc. attached to universities and colleges be helped to provide greater educational activities for the general public?

In what ways can universities and colleges be helped to form Centres where school teachers and others active in biological community development can go for in-service training and informal advisory contacts with the 'experts' who can help them in their work?

Can more research and development work in biological education related to community development be established in universities? What are the essential features of such work? Can we devise suggestions for arrangements by which inter-disciplinary studies can be undertaken? To what extent can such research and development be based on field work in communities? Can schemes be established between education and biological departments in universities in the more and less developed countries in order that specialist training in a 'developed' university be linked to community field work and university study in a country where the same level of expertise is not yet available? Would the establishment of comparative studies be valuable?

COMMENTS by R.A. Kille

The thorough analysis provided by the working document describes the problem before us and gives us a very good framework for our discussions. The document will direct our thoughts towards ways of approaching biological education for community development making use of existing educational institutions, and hopefully collaboration with these institutions by botanical gardens, zoological parks, museums, nature reserves, youth clubs etc.

I have no doubt that with the aid of the document expert participants will have little difficulty in devising curricula for community development, and many exciting strategies for making the programmes attractive and useful to the individual and the community. All I would say here is that I favour placing the emphasis in the first phase of these programmes on the development of the individual. I believe that from this egocentric approach will grow more naturally, powerfully and self-sustaining, the more altruistic community development.

Although confident that the meeting will succeed in fulfilling imaginatively many of the tasks set by the document, I have reservations. I have too often sat round a highly polished table in an exquisitely decorated and air-conditioned room discussing some educational problem with the converted, few, if any, of whom will be faced with putting the solutions into practice in the sands of the Sudan, the heat of India or the urban difficulties of Notting Hill. A major problem facing us in many aspects of community education will be to change the attitudes of educators in schools, teacher training colleges and above all tertiary institutes (especially universities). We will only tap the resources in schools, colleges and universities if we succeed in changing the attitudes of the majority of staff in these institutions. As far as the universities are concerned we will have to compete with the necessity for staff in these institutions to keep abreast of their subject and to maintain a high level of research which is acceptable to a highly competitive scientific community. Most biology teachers in school receive their biological training in universities and colleges and acquire their academic attitudes, so that there is resistance from biology teachers in school to teach socially relevant biological issues. Moreover, the universities and colleges ensure their teacher products do not escape from their academic attitude by demanding a high level of academic achievement in school leaving examinations which impose upon the school curriculum a highly academic approach, although only a very small proportion of the school population will go on to a university or a college. The teacher training colleges are caught up in this approach to biological education, knowing that their student-teachers have to fit into university-dominated school curricula and examination systems.

All participants will be familiar with this situation in developing as well as so-called developed countries, but the document hardly mentions strategies for changing attitudes.

I also feel that no matter how excellent the programmes and strategies devised for community education, it will require some kind of infrastructure to co-ordinate the various elements proposed in the document, and to initiate and maintain the momentum of activity by pupils, teachers, lecturers, parents, youth leaders and others. The kind of organisation best able to perform this function will differ from country to country, region to region. I feel however we should give some thought to this aspect of the operation so that some guidance will be available to those who require it. A good infrastructure will also add respectability to a community development programme and thereby more likely attract support from the tertiary institutes and schools. The drive and co-ordination might well come from a special unit established to co-ordinate the contributions from the various elements of the programme. These functions could very well be grafted on to teacher training colleges, as indeed they already are to a limited extent in some developing countries. In some cases it might even be the local universities, which, in spite of my strictures above, do sometimes have members of staff who are prepared to give their energies and expertise to the local community.

Any attempt to use tertiary institutes as bases for community education research and development will have to satisfy the usual criterion of scientific respectability. The best opportunities for using tertiary institutes will come from those institutes where a biological department is closely linked with education or where there is the possibility of setting up such a unit. There will also be some universities and colleges where a senior member of staff has opted out of the science rat-race and developed an interest in community development.

My reservations should not be misinterpreted. I firmly believe that an improvement in the biological contribution to community development must be made to enable more of the world's population to develop their full potential and lead happy and healthy lives. This is so important we must not fail. It is so important that we cannot rely on the school classroom alone. The adolescent and adult community must become involved in a continuing process of education so that any positive social biology learnt in the school is not diluted-out in after-school years in the adult world. We must plan carefully, testing each phase by piolot projects until there is sufficient experience and confidence to launch a major programme in one or two regions.

COMMENTS by E.B. Rugumayo

Scientific biology and community biology. The problem has been aptly framed: how to achieve the right balance between the three suggested approaches, i.e. putting emphasis on those biological topics which have implications for the community; placing a greater emphasis on human biology; and laying more stress on issue rather than knowledge studies.

My view on this matter is that we should regard this problem as a dynamic one, in which the three approaches are interrelated in such a way that each one of them in turn acts as the limiting factor at a point in time when efforts are directed towards the popularisation of biology education. The critical time when each of the three factors assumes the role of a limiting factor can not be hypothetically determined. It can only be arrived at as a result of concrete researches into the educational systems of different countries in general, and science and biology curricula in particular; and hence the need for studies which relate biological education to community development in individual countries.

In a country like Zambia where the traditions of scientific enquiry are simply not there and, strictly speaking, the present university and tertiary institution population is the first generation to be exposed to the traditions of the scientific method, due care has to be exercised as to which factor should receive the greatest emphasis. My view, which I am not in a position to support with the necessary arguments for lack of space, is that scientific and descriptive approaches should have priority, at school, over other approaches. On the other hand, at community level, issue studies should have priority over knowledge studies, and these should be combined in correct proportions.

Educational requirements for community development. Once again the problem has been appropriately posited: what would the people require from education in order that they can participate in development?

The dialectical relationship between the individual and community as far as biology education is concerned should be grasped. The actual learning process of biological principles and their social implications takes place only at a personal level, and because of the community. Thus creating awareness and developing the skills of social action are realisable at a personal level, manifested as personal development. The crucial question then is how does a community heighten its "awareness of what biological understanding is necessary for adequate development and demonstrate the extent to which it can lead to improvement in the quality of life"?

The opposite of this problem would be to substitute the individual for the community. A community does not become aware of the importance of biological education unless the training given to individuals places emphasis on the primacy of the social function of biology education. In other words, personal development should be seen as meaningful and possible through participation in relevant biological tasks which alone will ensure the integration of what has been learnt with what is practically feasible.

Schools and community education. This is a very controversial topic. Many developing countries have tended to place too much faith in the schools to the extent that these institutions have been overburdened with agricultural production units, military training and political education, in addition to the normal school academic load. The quality of many ot these activities is dubious; and the activities have tended to have adverse effects on the academic standards of the schools.

Whereas course modules may sound exciting, one can not see how these can prepare pupils for participation in community development when they leave school. The employment/unemployment patterns are such as to nullify most of the modular experience.

The proposals which stand out as more feasible include those which would be used primarily out-of-school. They would include those aimed at the adult population, and joint school community activities.

Tertiary education research and community development. It is quite apparent that universities and colleges in Third World countries are not doing their best in serving the community, more so in the area of biology education. Because of the flexibility and versatility of tertiary education courses, this should make it possible for some of these tertiary institutions to plan their programmes to serve the community. The proposals contained in this section are both relevant and timely.

Comments on Working Documents

COMMENTS by F.A. Gornall

I do not see any monumental stumbling blocks in defining the problems nor in drawing up a list of important topics nor in arranging these into priorities. However strategies for introducing innovations, especially innovations related to methodology will have to be promulgated by extreme bravery and profound wisdom. Did King Solomon agree to attend?

I am very strongly in favour of out-of-classroom learning and I will predict that much more of this will be promoted in school-related and non-school related programs of the future. Much of this learning will be associated with areas immediately adjacent to the school. It will reach out to nearby specialized education centres and will eventually embrace community-related educational programmes and projects. But whether the learning situation is within or away from the classroom I am much more concerned about teaching and instruction than I am of specific content. I fully support the acquisition of facts as a basis for higher levels of learning. In fact I would like every student to know absolutely everything possible about all aspects of his immediate and neighboring environments. I am also very much aware that traditional and too often current methods are largely inefficient in bringing us towards such a utopian educational achievement.

In the recent past the best intentions of curriculum developers in a number of programmes in a number of countries have had little impact towards the improvement in the teaching of the biological sciences because the methodology envisioned to support the pedagogy advocated has not been given the opportunity to assert itself. Some agents working against the acceptance of innovative programmes are:

a. Examinations: Final examinations still dictate the topics to be touched on in a biology course. Even with new programmes examinations continue to stress traditional topics in a traditional manner, "because they make good examination questions"!

b. Impatience: The time lag between the inception of any new programme and its true effect on the teaching of the subject is about one generation. Thus an impatience sets in because the world has not changed immediately after the programme was introduced. Shortcomings of students who had graduated prior to the new programme are blamed onto that new programme.

c. Inadequate Orientation: Education or re-education especially in respect to the methodology expected was in some cases completely lacking. The American BSCS biology course was prescribed for British Columbia by the Provincial Ministry of Education without any guidance or assistance. Consequently in most British Columbian classrooms the Yellow Version text was used as the course in the time-honoured traditional methodology of lecture, textbook reading and the filling in of blanks. The enquiry approach, the essence of the programme, never had a chance.

Perhaps the above points are worthy of consideration during our planning stages.

Further I believe that we should look at methodology from at least two different points of view:

a. For those who have been through the educational system and expect some form of didactic instruction together with those who were badly scarred from past schooling experience and may resist attempts at further education.

b. For those who have not been completely indoctrinated and are still receptive to innovative learning situations.

Another concern which might fit under the heading of methodology involves the local resource person. Again two aspects of this problem:

a. The "expert" brought from the community to the school to provide specific information on a community project. This is occasionally an enjoyable, satisfying and valuable experience. But more often an embarrassing, boring,fact-loaded dissertation unrelated to the needs or desires of the audience. These experts are well versed in their field, frequently volunteers, often giving freely of their own time, but totally lacking in methodology and unable to bridge the large gap between the presenting and the receiving of the information.

b. Many commercial organizations and community agencies when they become involved in education be it with schools or with public ecucation hire an "Education Officer". Usually this person will have a strong background in the subject area but again will be totally lacking in pedagogy and instructional techniques.

We do not want to lose the support or use of these people but we must find more efficient and effective ways of utilizing their expertise. Let's try for an emphasis on the education side of Biological Education.

An anecdote which seems to fit about here. At a recent meeting of the directorate of a local forestry group the education chairman described in glowing terms the numerous charts that his group had prepared and distributed free to schools:

Chairman: "and before we sent them out we had them checked by professional foresters."

Question: "Did you have them checked by professional educators?"

Chairman: "No" ... !

I favour information and education centres on all and various community projects. The larger the better. But even temporary units such as a kiosk in a shopping centre and staffed by students can be very successful. We have found them most effective, for example, in providing both data and publicity about the proposed Tynehead Zoological Garden in Vancouver.

COMMENTS by H. Bergmann

Introduction. This paper attempts to define the basic terms of community and community development, goes on to examine some of the implicit hypotheses on which many of the recommendations in the

Working Document seem to be based, and suggests alternative ways to have biological education contribute to community development.

Where this paper refers to Third World Countries it builds on experience gained in Africa south of the Sahara.

Community. Since "community development" is the key term for discussion, it might be necessary to define what is meant by community. Community as used in the context of community development refers to a settlement of people:

"Community is ambiguously defined alternatively by co-residence within ecologically defined boundaries or by the intensity of group cohesion." (Benini, A., 1975, p. 22) [1].

Further precision can be got from the following definition:

"By community we mean a group of people, often but not always spatially close together, who recognize each other as members of a group larger than the extended family and who recognize a common decision-making body..." (Bergmann, H., and Bude, U. 1976, pp. 122-3) [2].

Especially in the context of the community development doctrine, the meaning of community is often further restricted to the small rural community with very strong overtones of harmony, solidarity, and co-operativeness. The rifts and tensions tearing apart small rural communities are easily glossed over.

Since education in general, and therefore biological education, too, is nationwide, limiting ourselves to the small rural community would be wrong. Problems of the urban community would need to be included, too.

The term community is also sometimes defined as a biological system. If all systems made up of living organisms are by definition biological systems, then the statement is true. However, to the non-biologist it sounds a bit surprising. He usually thinks of communities as socio-political and sociocultural systems.

Community is, for all practical purposes, the lowest level of the hierarchy of territorial administration. It is at this level that most schools and other educational institutions are localised. Practically all primary schools cater for one or a very limited number of communities. This is where the idea of educational action directed towards one community in order to foster its development seems most plausible and appealing.

Secondary schools and most institutions of non-formal education are located in a given community but serve a much larger area. If biological education is supposed to further community development in this broader context, the practical meaning - who does what, with

1) Benini, A.A. Community Development in a multi-ethnic society, unpublished diploma thesis, Bielefeld 1975

2) Bergmann, H., and Bude, U. An Analysis of Existing School-Community. Participation in a Central African Country, in: King. (Edit), Education and Community in Africa, University of Edinburgh, 1976, pp. 121-160.

and for whom - becomes much less straightforward.

Community Development. Since the first large-scale community development programmes were launched in India and Pakistan, the term has become something like a trade mark. It denotes an approach to integrated development as opposed to sectoral development and carries a strong emphasis on "development from below". The members of a community as defined above pool their efforts and resources in order to achieve common goals. Important in this respect is the idea of a community acting as a unit and speaking with one voice.

The term has been officially defined by the United Nations and means to them a process

> "whereby the efforts of the people themselves are united with those of governmental authorities to improve the economic, social and cultural conditions of communities, to integrate these communities into the life of the nation, and to enable them fully to contribute to national progress.
>
> This complex of process thus consists of two essential elements:
> - the participation of people themselves in the measures taken to improve their conditions by utilizing as far as possible their own initiative, and
> - the provision of technical and other services in a form that encourages and makes effective spontaneous self-help and mutual aid." (UN 1959, quoted in Benini, A., 1975, p. 1) [1)]

In many instances, community development programmes have degenerated into "social activities" (e.g. needlework for women) without any immediate prospects of improving the lot of the ordinary community member, and state administered programmes of enlisting community members for unpaid work contributions. It is necessary to clarify our understanding of community development with special reference to topics such as —

a) who defines the aims and operational targets of development activities at community level?

b) who decides on the actual programmes?

c) what is the time horizon of the programmes?

d) who are the main beneficiaries of the programmes at community level and beyond that level?

e) who is going to contribute what to the programmes?

Only then is it possible to discuss how education in general can contribute to community development, and what the place of biological education might be in such an endeavour.

Education. We can take the meaning of education as used in the IUBS-document for granted, keeping in mind the distinction between formal, non-formal, and informal education. However, any realistic strategy will have to place its emphasis on formal education. Formal education is better established than any other form of education and

[1)] Benini, A. (1975) loc cit

can be counted upon for long-lasting actualities whereas most programmes of non-formal education either remain at the pilot stage, or reach a very limited audience, or are very limited in time (e.g. alphabetisation campaigns and adult literacy courses).

Old myths never die. It would seem to me that the suggestion contained in the working-document incorporate ideas and proposals that are not at all specific to biological education but are the stock-in-trade of the international experts on educational reform, and which, to date, have not proven their feasibility but look very progressive and 'off-the-beaten-track'.

I should like to argue that to base projects in biological education on such ideas will almost certainly doom the projects to failure. The ideas referred to are -

a) the possibility of a closer link between education and community, in practice between school and community;

b) the beneficial effect of such a close integration on the community,

c) that community members would turn to the schools for leadership in community activities;

d) the activities of the school can be easily enlarged and to incorporate development tasks;

e) the role of the teacher can be changed to incorporate community development elements;

f) the possibility of achieving steady and reliable progress in learning and coping with development projects at the same time;

g) the possibility of getting non-school personnel into the schools for teaching assignments on a regular basis.

Let us look at these ideas in some detail. If they are correct, then the desired effect of biological education on community development is possible. If they turn out to be unrealistic, then the aims of biological education, as far as community development is concerned, would have to be pursued by different means.

a) The links between school and community. In one sense the links between school and community are already surprisingly close. Out of sheer necessity, community members contribute much to the running of the schools. This they do not because they expect an immediate impact of education on the local situation but because they want their children to compete successfully for paid jobs. They then expect contributions by the successful ones in local affairs. But the results are rarely what the development planners would like them to be.

Traditionally schools and other educational institutions have been accepted as separate entities which pursue programmes not at all related to current community affairs. This is as true as much for coranic schools and initiation training in the "bush schools", e.g. the Poro in West Africa as far modern primary schools. The content of education is far from the realities and problems of the situation in the community. Knowledge and skills transmitted by schools tend to be so divorced from everyday realities that they become next to useless in relation to community development except that they pave the way to employment. Closer integration appears to be possible only when this fact is accepted.

Practical integration becomes possible with devices such as the use of school premises for community purposes. This depends largely on good relationships between headteachers and the community. A most appealing idea would be to make teachers leaders in community development projects. We shall come back to this later.

b) The benefits of closer links between school and community. It is often said that communities can not but gain from closer links with schools. It might be worthwile examining in some detail some of the ways. It assumed such benefits would be obtained.

It is assumed that schools teach better ways of doing things. This assertion begs the question. Apart from the value judgments involved it can be shown, that in the past, schools taught wrong content (Bergmann and Bude, 1978 [1]), Bergmann, 1978 [2]). Unless really thorough knowledge about the local ways of going about things is available, whatever improvements are offered by schools or other educational agencies are likely to be only partially relevant.

It is also assumed that adults learn from children. The idea is that pupils learn improved techniques or new types of behaviour in school. They are asked to put these into practice at home. As they demonstrate the superiority of the new ways e.g. in hygiene, nutrition or farming, other household members will follow their example. The ideas rests on three hypotheses. The first is that pupils are able, by simple practice, to demonstrate the superiority of the new ways. However, assuming that there is some real superiority, this will by no means be evident immediately since the effects of cleanliness, better food habits, and improved farming methods, for example, take a long time to become visible.

The second, alternative hypothesis, is that schools have such a high standing that whatever they teach is accepted. This is certainly not true in contemporary Africa and probably has never been true, in general, in the past. Schools have not been valued for the practical relevance of what they teach.

Thirdly, it is hypothesised, that the average adult is willing to learn from children in important domains of behaviour. This is not true in any of the developed societies. One wonders whether it is more than a rare coincidence in societies conferring very high status on old age as is the case in Africa and much of Asia.

1) Bergmann, H. and Bude, U. Thesen zum Thema Grunderziehung und Dorfentwicklung. Paper presented at the regional seminar on agricultural extension, Victoria, Cameroon, November 1978.

2) Bergmann, H. Agricultural Instruction in Primary Schools. A contribution to Rural Development. Development and cooperation. No. 2, 9-11, 1978

c) Community members would turn to the schools for leadership in community activities. Usually, communities do have a structure for problem-solving and decision-making. Problems that they define as urgent they will try to tackle themselves. They might lack technical know-how, but it is by no means certain that they would approach the teachers for help. Teachers are very often approached for personal matters, and often they are asked to participate in committee work because they can write and have relations with the administrative bodies and know their way around. The school does not fill a vacuum in terms of leadership. Any move to make schools assume a leadership position in community development may well meet resistance and be perceived as competition rather than a necessary service.

The idea of the school leading the community makes sense, however, in a different context. If community development programmes are planned and executed from above without the communities participating in the decision-making, the problem no longer is to help a local community to carry out its own development projects but rather to make sure that the required local efforts are furnished at the right time and in sufficient quantity. It means getting instructions to the community members and controlling their implementation. This can be done by mobilizing the schools. After all, the school is the public institution most widely implanted. In such a context, the school becomes an arm of the central development bureaucracies without any real leadership function.

d) The activities of the school can easily be enlarged to incorporate development tasks. This is a technocratic view of school and, in fact, of any task-oriented organisation. It tends to forget that the task specifications made by those who are in formal control over the schools are only one of several factors determining the actual programmes and actions of an institution. The fact that someone decides to add a community development element to the task definition of schools does not ensure that substantial activities will follow. The traditional expectations of those directly served by the school are also important. They may lead to lack of enthusiasm, the refusal to co-operate, maybe be to resistance against imposed extra activities. There are many examples of lip-service given to such ideas as "the Green Revolution" or "Community Work" without activities resulting in schools, to illustrate this point. Formal compliance with instructions at a minimum level is the typical reaction to the redefinition of tasks that go beyond long-established traditions.

e) The role of the teacher can be changed to incorporate community development elements. The myth of the teacher being an innovator has a long tradition. In fact teachers rarely are innovators except in small pilot schemes with their specific conditions of motivation, guidance, and close supervision. The myth assumes that the traditional task of the teacher has been carried out satisfactorily, or that it has lost importance or that it does not utilize his capacity to the full.

In many countries, none of these assumptions is justified. Many teachers are poorly trained and cope with the task of teaching the three R's with some difficulty. Changing the role of the teacher would mean to add to his work load. If his salary can be raised accordingly this might be feasible. But often a new task means making him work more for the same pay.

Also teachers usually know the pay scales, the conditions of work and the possibilities for extra income in administrative and other jobs. One cannot assume that they are willing to accept additional tasks unless others are asked to do the same.

Apart from this, the teachers' role as a community worker and innovator would need a clear definition of the task involved; a specific training different from the usual teacher training, well- defined access to the competent administrative services, and extra material resources.

f) Achieving steady and reliable progress in learning and coping with development projects at the same time. The working document advocates a project approach to biology teaching. Communities, it is said, should define problems on which schools would work. Again, there seem to be some difficulties.

There are, for example, few community development problems related to biology and truly identified by community members, not by someone from outside, that schools can tackle successfully. These problem, e.g. in hygiene, child care, water supply etc., cannot be solved in a short time. They lead to specialization and prevent broad learning. The range of community development projects which can be started at anyone time is limited. This, in turn, limits the choice of teaching topics and precludes learning on a broad basis. Also, beyond the most basic activities, community projects can quickly require technical expertise beyond that of teachers.

The project approach becomes feasible if the ambition to directly contribute to community development is abandonned. Instead, teachers can identify local problems with a biological component. If adequately trained and backed up by educational materials they can devise teaching/learning projects that are limited in scope as to topics, resources, and time. This would reconcile the range of interest of the pupils, the requirements of national syllabuses, and the potential of primary and secondary schools for action.

g) Liaison with other people. The recommendations also mention that biology teachers would have to liaise with people working on similar problems, e.g. the health worker, the agricultural extension officer, the veterinarian; and that these people should be involved to some extent in school work. It is an illusion to believe that this can be done beyond an informal level. The various specialists get their instruction from their respective employers, and their duty schedules usually do not allow them to participate regularly in school activities.

Furthermore, they are not available everywhere so that a general strategy linking school and community cannot rely on them. Also only a few people would be willing to teach without financial rewards.

The contribution of biological education to community development. Talking about community development in general is a simplification which precludes any operationalization of projects. Communities differ, and so do community development problems. I suggest we need to distinguish a number of community types. For a start I should like to propose a classification according to two dichotomies: rural/urban and developing countries/industrialized countries:

	developing country	industrialized country
rural	the small rural community	the "urbanized" countryside community
urban	spontaneous settlements, residential segregation within communities	industrial/administrative macro-centres

It is evident that the community needs are different in quality and intensity for each of these situations; and that the contribution of biological education will be different, ranging from direct action at the level of the small rural community to small scale political action in the industrial and administrative macro-centres. In a small rural community, the teacher could end a teaching sequence on drinking water by building a series of sand filters near houses. In the huge urban communities of the industrialized countries teaching about the purification of water could lead to the writing of a petition in order to obtain drinking water of good quality.

For practical purposes, such as curriculum planning in a given country, it is likely that one would have to refine the classification further. Thus, a study in Cameroon identified six community types ranging from isolated rural communities, accessible only on foot, to urban and semi-urban communities. It was possible to spell out a few topics for joint school-community projects in agriculture, hygiene, and health education according to community types (IPAR-Buea, 1977 [1]).

The learners' relationship to living things. Biological education would have to take into account the way children experience living things. It makes a difference whether children get to know plants mainly for ornamental use in gardens, parks, flower pots, etc., or as building material. It will also make a difference whether they see plant-food growing or whether they know most of it only in the form of processed products. In relation to animals, it will make a difference whether children have a first-hand experience of wild animals, go fishing, and are familiar with domestic animals or whether they know them mainly from books, films, the zoo, as pet animals or as pests. Differences in this respect, but also with respect to direct exposure to many facts of human life (birth, sickness, death) and to abstract and theoretical biological knowledge are closely related to

[1] IPAR-Buea Report on the Reform of Primary Education, Buea, April 1977

the community types outlined above. The same is true with regards to ideas about causality, and about the emotional value of living things. The source of first-hand experience with living things is the learner's immediate environment, the source of knowledge of a more abstract type and of values relating to life in his community's culture.

I should think that any attempt to relate biological education to the realities of a community would include the following sequence of steps:

a) Compiling knowledge about the socio-cultural representation of biology in the community.

b) Evaluating this knowledge in the light of accepted and tested biological knowledge,

c) Deciding on how to handle serious contradictions and divergences,

d) Preparing instructional material aimed at correcting gross errors in the local socio-cultural "biology".

This would lead to curricula that are related to realities at the community level but without any ambition to achieve immediate tangible results. The aim would rather be to prepare the ground for the action of specialist teams whose programmes and activities are based on modern biology, its assumptions and knowledge of cause and effect.

Identifying problem areas for community development. Problems related to biology that have emerged relatively recently as issues in the large urban centres of the industrialized world cannot be projected on Third World countries. Pollution, for example, is a problem in many urban areas in the Third World, but in a different way from that in the cities of the industrialized countries. It is no problem in rural communities. Conservation and recycling are meaningless in communities where materials are carefully used that elsewhere would be regarded as litter. An the approach to demography, family planning, and health education must be radically different in communities threatened by extinction through emigration compared with overcrowded urban areas.

It would seem most important to find out each community's biology-related problems and to take its peoples' way of identifying and presenting them seriously instead of assuming some sort of universal set of problems for which solutions can be devised in remote offices. It will need the humility to patiently sit with, and learn from, community representatives who often are not very articulate and who often have learnt not to trust people from outside who come and promise yet another bright future.

Applied biology. There is no doubt that applied biology can be usefully included in education. Most benefits will accrue to individuals - farmers who get better yields or an easier job, people remaining healthy, mothers experiencing fewer pregnancies but more healthy growing children - rather than to communities as a whole. But, in turn, these effects will contribute to the overall development of communities.

Certainly, basic scientific procedures and thinking can be taught through applied science better than by a pure science approach because the link with everyday life is more attractive. In my own activities, devising a curriculum for primary school agriculture in Cameroon, the following steps were taken

a) Information about local crops and farming techniques was collected.

b) The knowledge was documented for schools.

c) Ways by which schools can continue to collect such knowledge were proposed. (Here is a possible contribution to community development which does not exceed the normal task of the schools.)

d) Teaching/learning sequences that relate classroom work systematically to the agricultural calendar were devised.

e) Instruments for simple observation work were designed.

f) Series of simple experiments to be conducted on the school farm were proposed.

g) Information about recent developments in agriculture were made available. In our case it was information about eco-farming and related approaches to tropical agriculture which fit in with traditional farming. This represented another contribution to improvement at community level since it opened the door to progress along lines of action and thought familiar to the local population. Again, it does not mean an extension of the role of the school or the teacher but simply better inputs for them in usual tasks.

BIOLOGY AND POLITICS
by M. Atchia

Psycho-social evolution is now so rapid that, for the first-time in the history of mankind, it appears that we do not know where we are going or are heading straight towards self-destruction.

Phrases like the above are not uncommon these days: we have heard them from Commoner and Dorst, from the preamble at Stockholm, from the two Club of Rome reports (particularly from their second report) and above all from Goldsmith et al in the Blueprint for Survival. An underlying assumption in all the above is that psycho-social or human evolution is a conscious process and is largely directed by man himself, unlike organic evolution which is essentially a response to changing envionmental conditions. Thus, there is the need to define clearly where we are and where we are going - "we" refers here to both the human species and to the ecosphere. It is in this context that I want to analyse briefly policy-decision making in Africa (and elsewhere) and to point to the relevance of basic biological principles to that process. I am not so much referring to politics as it is practised today by most Western countries and, to varying extents, by many African states, but to politics in its more revolutionary and autocratic form. The former is the art of regulating the day-to-day and year-to-year life of societies in their complex structures, while the latter may try to redefine goals, destroy societal structures and attempt to establish new ones. Worse still, politics in some African states has set itself the task of defining or designing new principles on which society is to be fashioned, for example the Ujaama of Tanzania, the Fokolona of Madagascar, the Uhuru spirit of Kenya and a wide variety of marxism, such as those of Mozambique, Angola, Ethiopia, Seychelles. I have made a value judgement here in using the term "worse still" - not in any way a reflection on the comparative values of regimes nor an absence of recognition that trying to design or discover new principles is, in a way, scientific research, but because decisions under some of the regimes where "things have been re-defined" are often taken against basic human (i.e. biological) principles. Hence the relevance of biological education, not just for the political class but for the masses whose lot it is to follow or reject experiments of the Campuchean or Guinean (Bissau) nature. T. de Chardin [1] stated a paramount biological principle where he recognised that psychosocial evolution does not necessarily follow the same principles as organic evolution, where, for example, in the classical case of struggle for existence one group eliminates its rival and survives.

Graham-Smith [2] argued that within the biosphere the basic structural unit, the cell does not operate in the repetitive way in which elementary particles and atoms do in forming the universe, but "go through a constant state of fundamental change", with compensatory mechanisms to keep to alive. Were a cell to be denied energetic and materialistic interactions with the environment it would cease to function. A parallel can be drawn between the cell and its environmental interactions and the human species as a psycho-socio-organic unit and the biological principles necessary for it to survive and progress.

Thus how important is human population control? And what to say of the Government of a densely populated country which rejected population control? How biologically necessary is inheritance? Physical inheritance is, of course, fundamental to all living things. Is not cultural inheritance, the transmission of modes of life and knowledge as important to human societies? How then to view the decision of a state to abolish the right of inheritance? Is the right to transmit property and belongings, particularly land and house (the niche constructed by the parents and passed on to the children) a requisite of progress? What relationship is there, if any, with the concept and importance of territoriality?

The importance of the forests of Amazonia and the Congo basin to the CO_2/O_2 balance on earth cannot be denied, biologically. Would political decisions which will result in the partial destruction of these forests have been taken had the decision-makers a high degree of biological education?

---

[1] De Chardin, P.T., Le Phénomène Humain. Editions du Seuil, Paris, 1955.

[2] Graham-Smith, W., Organization in natural systems. The Ecologist Quarterly, 2, 1978.

What about the value of cross-breeding as contrasted to in-breeding and the various tribal, racist and other segregationist policies prevalent today? And what about the genetic dangers of atomic radiation?

The principle of rational utilization of a renewable resource is so simple that a 10/12 year old can understand it. And yet some members of the political class, through complex semantic wranglings, can show you why it is valid to exploit "their" natural resources immediately until complete depletion, if need be.

How important therefore it is to convey the biological nature and needs of man as well as the rules which govern this planet and its constituent spheres to everyone but, particularly, to those who make policy and take decisions.

THE PLACE OF TECHNOLOGY IN BIOLOGY CURRICULA
by W.H. Dowdeswell

In Britain, the 1960's ushered in a period of unprecedented change in curricula of all kinds particularly those relating to the teaching of science at the secondary level. But there are now signs that the period of innovation has run its course; indeed it may well be that science teachers are suffering from an excess of novelty. One of the most widespread criticisms of present-day science teaching is that while syllabuses adequately reflect modern scientific thinking and teachers show an increasing awareness of the need to present the subject in a scientific manner, nonetheless the outcome tends to be lacking in humanity. This could be construed as implying that insufficient emphasis tends to be given to the presentation of science as a branch of human endeavour; we stress the product at the expense of the producer. An alternative interpretation could be that too little attention is paid to the place of science in society particularly in the realm of technology. This might well account, at least in part, for the antagonism to science among a significant proportion of our school pupils. Science, like all other branches of human knowledge, is firmly grounded in the needs and thinking of society. To what extent, we may ask, are such links reflected in the kinds of teaching to which our students are subjected today?

Modern science curricula at all levels pose a curious paradox. For while it is universally acknowledged that human society is largely dependent upon science in the form of modern technology, nonetheless the study of technology as such hardly features in courses at any level. By technology is implied not just the applications of science for human needs but also a consideration of the economic, social, ethical and aesthetic contexts in which those applications are made. This argument applies as much to biology as to any other branch of science.

If such curricular changes are to be brought about there are a number of questions that will need to be answered. Some of the more immediate are,

a) if new material is to be included in curricula what should be omitted? At all costs we must avoid the sin of growth by accretion.

b) What is the availability of suitable teaching material? The immediate answer appears to be that although a vast amount of information is to be found on the files of biology - based industries, in research establishments and elsewhere, relatively little is in a form in which it could be used for teaching without considerable modification.

c) Have our science teachers the techniques for teaching non-science material? Such experience as is available suggests that science teachers do not adapt readily to teaching in a more humanist vein. This is only to be expected if we accept the premise that the teaching of science is likely to appeal most readily to convergent thinkers while the humanities are more likely to attract divergent thinkers.

An important component of technology is decision-making. Experience with projects such as the Ridpest File [1](a set of decision-making exercises based on the use of pesticides) has shown that problems such as those enumerated above can be overcome to varying extents.

THE STRUCTURE OF BIOLOGICAL EDUCATION

by P. J. Kelly

The term biology tends to denote a concise body of knowledge and activity. It is depicted as one of the sciences. I am concerned with something much more encompassing: the study of life focussed on the nature of living beings (including ourselves) and the relationships between their form, functioning and environment, both in the present and over time. In this sense we are including biochemical, physiological and anatomical aspects on the one hand and, on the other, behavioural, ecological and cultural dimensions. This does constitute a constitute a concise body of knowledge. It certainly is not one science and, indeed, can be said to go beyond science. It is a mighty sweep of consideration and sorting out its meaning poses one of the major problems in biological education.

The fifties and sixties were times of considerable upheaval and extension of biological knowledge. At one end of its spectrum were the discoveries of the molecular biologists, epitomised by the elucidation of the structure of DNA. At the other end Lorenz and Tinbergen opened up the field of ethology. And there was much going on in between. This stimulated much thought about the content of biological education and it was to the advantage of curriculum developers in this respect that they were invariably able to start from scratch in designing their new courses.

In order to gain some idea of the variety of thinking which occured,

1) Dowdeswell, W.H. and Wells, J. The Ridpest File Journal of Biological Education 11, 53-58, 1977.

let me focus on a simple scene which I observe daily around Richmond Park. It occurs every morning and evening: a large-scale exodus of dogs taking humans for walks and invariably spending periods of time by trees. Surely it is worthy of biological study; but how could we organize it?

The easy way out is to make it a specialisation: call it, say, biopedestrianology and soon Ph.D.s, papers and possibly even a best-seller may well accrue. In terms of general education, however, we have to ask how does it fit into the total design of a biology course and then we come up with a variety of answers.

We may argue that if a biology course is attempting to portray the insights of biologists then it should be structured in terms of research perspectives. Thus we would divide our study of the park scene into the biochemistry of ......, the evolution of ......, the ecology of ..., the ethology of ...... and so on. As a result, a variety of disconnected images of our scene would be presented in the course. For example, of chemicals and laboratories, of fossils, of food web diagrams and mathematical formulations, and of behaviour signals and responses, respectively.

Another alternative would be to divide up the course according to the types of organisms within the scene. Thus we would have the zoology of the park scene, the botany of ......, the human biology of ......, the microbiology of ...... etc.

You could also conceive a course based on levels of biological organization. Then, for example, we could first look at the molecular and cellular aspects of the organisms in the scene, followed by their tissues and organs, their whole bodies and finally study populations of the organisms.

My last example of a possible structure for course design, but by no means the last suggested, requires us to stand back from the park scene and ask what are the major themes or principles of life? What have living phenomena in common? This may well lead us to use ideas, which we will call themes or principles, such as the relationship between structure and function, and organsims and their environment, as a framework for study. Continuity over time (heredity and evolution) and communication (applied to molecular processes as well as the behaviour of whole organisms) are among others which may be included. Such schemes have been variously used to construct courses. There is clearly overlap between them but, especially, we can draw a distinction between the 'Research', 'Levels' and 'Principles' categories which are to some measure theoretical constraints and those of the 'Types' set, which is concerned with more discernable phenomena. The type system was for years the traditional form of course design in biology. The new developments tended to move away from this but then exposed a fault in the alternative, more abstract constructions which were invented. These are concerned not so much with living things which you study but ideas you draw from studying them. Thus they can be used to provide outlines for constructing courses and phrases you can use for summing up at the end of courses. However, they have limited value in terms of defining what you actually study. I suspect that one of the lessons we have learned from the sixties is the limitations of such theoretical schemes and we must look for something more concrete.

One of the reasons why it is not easy to devise schemes for biological education is the intriguing paradox that within the make-up of living beings there are both similarities and differences. The problem is that we find it difficult to accept that these apparent protagonists are inevitable bedfellows.

To illustrate what I mean may I refer you to a conversation in the columns of Nature in the early sixties. It was the time when, with the major successes in molecular biology, microbial genetics and biochemistry, there was a strong move towards reductionist opinions, the idea that the form and function of living material was explicable in physico-chemical terms and that, in particular, as Professor D.G. Catcheside put it "The reality of biology as a subject lies in the unity inherent in the apparent universal similarity in the constitution of cells."

Now, that there are significant similarities between the cells of organisms is not disputed but there is a variety elsewhere in the biological world and Mr. W.B. Yapp contested Catcheside with this statement. It is particularly related to the idea that - based on the type of philosophy advocated by Catcheside - the division of botany and zoology should be merged in a single subject biology as, in fact, happened, for example, in JMB A Level examinations, and to some extent in universities.

Yapp said "...what is important is that for me and many others the interest of biology lies in its diversity. To say that because sperm tails and fern cilia have a similar structure there is a 'lack of serious distinction between plant and animal material' is ludicrous. Professor Catcheside is not an oak tree nor am I a nettle, and to pretend that we are is nonsense. The 'universal similarity' (which every schoolboy knows to be extremely limited) makes the diversity more, not less, remarkable."

Catcheside is a microbiologist, Yapp a zoologist. Thus it can be said that their views reflect their interests, but this does not make them less plausible. They are, in fact, both right. Cells are similar, organisms vary: both have to be taken into account in devising courses. Such polarisation of views does not help in itself with providing a structure for biological education. However, I feel there is hope in this controversy and, interestingly, in the misuse of the term unity when what was meant was similarity.

Thus, when Catcheside talked about fern cilia and sperm tails exhibiting unity, he really meant that they were similar. Fern cilia and sperm tails have no relationship in the sense that they unite; they rarely meet, they do not do anything together. However, if you take Yapp's oak tree and nettle and, even, Catcheside's micro-organisms, they clearly have many differences but at the same time can be intimately related. Within an oakwood, where there is loose soil and decaying organic debris derived from leaf fall, the stinging nettle forms localised societies often to the exclusion of other herbaceous species. Micro-organisms form an integral component of soil life and are responsible for decaying the leaves that allow the nettles to thrive. In other words, there is a unity of supportive ecological relationships. It represents a synthesis of the variety of organisms into a functioning whole.

There is a similarity between this description of the oakwood and our scene in Richmond Park. Both are syntheses of a variety of phenomena. Yet each of the schemes (including the 'type' system) I outlined for studying the park scene would result in a portrayal of only parts of the scene. None would depict the whole. Yet that is what we were supposed to be studying. The schemes' deficiencies lie in their inhability to synthesis ideas into the reality of the scene itself.

I would like to suggest that this idea of synthesis is central to the study of life phenomena. Through it we can incorporate and merge the contesting ideas of other formulations - types, levels, principles etc. It recognises that the essence of living entities is holistic rather than reductionist.

There is already movement in the direction of using a framework of syntheses for biology courses, putting an emphasis on organisms as individuals and communities. I certainly see it as the most relevant one to pursue at this time.

The Biosocial Dimension

As we have seen, the scope of biological education extends from its links with the physical sciences in one direction to psychological and social aspects in others. During the 1960's the physical sciences were drawn into biology courses but more recently attempts have been made to provide a social dimension as well, particularly in human biology. It has led to the definition of a category of phenomena which are described as biosocial.

The term biosocial is not synonymous with social implications. Many biological phenomena have social implications. The effect of vitamin C on the body has social implications in terms of the provision of food but this I would not describe as biosocial. On the other hand, feeding habits do constitute a biosocial phenomena. They can, of course, also have social implications.

A biosocial phenomenon is one which has roots in the physical makeup of an organism but which is expressed in a social context. The study of it brings biology - particularly human biology - into liaison with aspects of sociology, psychology and anthropology.

There has been a sad history of marriage and divorce between biologists and social scientists. The liaison in the Victorian years of Social Darwinism was impetuous and to a great measure ideological. It was a social scientist, T.R. Malthus, who provided the final clue to the theory of organic evolution through natural selection conceived by the biologists Charles Darwin and Alfred Russel Wallace. This, in turn, was used by Herbert Spencer and other social scientists to depict social processes in a naturalistic and evolutionary framework so that by 1909 it was said that "sociology, anthropology and the comparative study of religions were predominantly evolutionary sciences political science was under strong Darwinian influence, some economists were asking in a worried way why their subject was not 'evolutionary', and the new subject of eugenics was at its most fashionable".

As the twentieth century progressed, the partnership divided. Biologists, accepting the theory of evolution, confined themselves more and more to the laboratory, fathoming out the physical basis of life. The social scientists, particularly the sociologists, became less concerned with social evolution and turned more to the study of the structure and function of institutions. The interests of the partners moved apart, and the process was exaggerated as both sought to formulate a respectable identity; biology through an association with the physical sciences, social science by not 'whoring after theories drawn from natural science' and by acquiring an autonomous, academic responsibility and respectability.

Today, after this lengthy estrangement, there is, at least, partial reconciliation. It has come from changes in interests and emphasis of study rather than new discoveries or new integrative theories. There is also mutual recognition that certain topics cannot be adequately studied without co-operative endeavour. Several examples of this rapprochment could be cited but two will suffice, one of social scientists looking to biology, the other depicting the reverse trend.

The study of the social structure of industrial societies has led sociologists to the recognition and description of social stratification and from there to the study of social-mobility - the movement of people from one economic and /or cultural stratum of society to another, either within a generation or between generations. Sociological and demographic interest in the structure and growth of the human population both stimulated and contributed to this trend; and the efforts of educational psychologists to investigate the distribution of ability in the population, although at first parallel to the work of the sociologists, eventually merged with it. We can see the later years of this trend exemplified in a line of events in British research which commenced with the Royal Commission on Population reporting in 1949 and the notable study of social mobility by the sociologist David Glass and his colleagues at the London School of Economics, published in 1954. Questions about movement between social classes eventually led to those concerned with human ability and reopened the issue of the relation between social class and I.Q. studies earlier. As this research proceeded, it was inevitable that questions with a biological component would be posed. For example: To what extent is ability inherited or acquired? What influence does differential mobility between social classes have on the distribution of ability? Currently, such questions are to some extent stimulating cooperative research between social scientists and biologists. Although, judging from the public controversy it throws up, the extent to which a marriage is in sight is not clear. But, in less public view, the work of Halsey, Glass, Thoday and others has made significant strides.

Biologists have looked more to the social sciences because of their growing interest in the human species. 1964 saw the first significant University-level text book on Human Biology published in Britain. It defines human biology as "not so much a discipline as a certain attitude of mind towards the most interesting and important of animals. Human biology portrays mankind on the canvas that serves also for other living things".

Interestingly, the text confined itself to human evolution, genetics, biological variation, growth and ecology. It largely omits those areas most allied to the social sciences despite the contention - in the Foreword - that "Human biology deals also with human behaviour - not with its wayward variations from one individual to another, but rather with the history and significance of, for example, family life; or of love, play, showing off, and real or sham aggression. Finally, and most important - because most distinctively human - it must expound and explain the nature, origin, and development of communication between human beings and the non-genetical system of heredity founded upon it".

Other biologists have not been so reluctant to tackle these areas. For example, ethologists such as Lorenz and Morris have trod controversial paths attempting to identify the non-human roots of human behaviour. With even greater boldness, Sir Cyril Darlington has interpreted human history as a product of genetical and mating systems, and in Professor J.Z. Young's 'An Introduction to the Study of Man' the attempt is made to formulate a broad biology of man 'tracing the sources of human activity from their biochemical basis to the highest levels of consciousness'.

Some of these approaches have certainly hit the headlines. Some have been dubbed the 'New Biology' and have led to considerable speculation. In 1972 Antony Jay, a former BBC producer, wrote about Homo sapiens corporalis - 'Corporation Man'. He says "What made the pieces of the corporation puzzle come together for me was a sudden relevation about the evolutionary, biological nature of the production team. The New Biology shows beyond all argument that the group, so far from being an accident of corporate organisation, is one of the most ancient characteristics of our species". As so often is the case, the popular image distorts both the value and the substance of the scientific work underlying it.

The distortion has made the relationship between the social and biological sciences an uneasy one but, whithin its limited scale, it has today acquired a firmer foundation. One may, in fact, argue that it is inevitable. If we are to understand ourselves and our own species we have to understand the links within our biological and cultural natures. To be a human being is to know they are not separate phenomena. Separation and particularly the antagonisms of polarised argument are merely artefacts of sub-cultures of scholars and ideologies. These may be characteristics of human knowledge. They should not be misinterpreted as knowledge of humans.

It is, I feel, one of the great challenges to biological education to formulate a biosocial synthesis in a way that gives it credibility and a rightful place in general education. As the challenge of the sixties was to unite biology with the physical sciences, so in the coming years I suggest we have an equal task - to unite with the social sciences. Why do I feel this is so important. First, because it is there: we can't ignore it (although some would wish it) and it won't go away. The content of this biosocial dimension goes to the basis of so much of human concern; the meaning of ourselves in the world and the perpetual riddles of the relation between our physical and cultural selves seen in health, intelligence, sex, population, race and like subjects.

Secondly, it adds an important process dimension to biological education. Physical science-based biology is one in which we do experiments, in which we are concerned with how the body works, and of what chemicals it is composed. It is the same approach as one adopts for the study of inanimate material. However, as John Pringle pointed out sometime ago, it is only one of the two biologies. In the biosocial sphere we ask questions of a different sort: 'What is the function of this or that feature of man or of society? 'How and why did this feature come to be what it is?' These are questions we rarely ask about the inanimate world. Also the study of biosocial phenomena does not lend itself to experiment, at least not the conventional types. Other methods are required for study.

By introducing a biosocial dimension to biological education we add to its wealth of perspective and enquiry. We will learn something more of the many ways we have to look at life.

Knowledge and Issue Studies

The debate on the structure of biology is essentially an in-house discussion amongst biologists and biology teachers. During the end of the sixties and the beginning of this decade other matters got more than their fair share of public debate. I have in mind ecology. As science slipped from its pedestal the environment jumped on to it. You will remember there were Conservation Years; Paul Erlich of the United States warned us of the 'Population Bomb'; the journal 'Ecologist' was established and brought out its 'Blueprint for Survival'. There was Torry Canyon and one of the quickest produced scientific reports ever. There was 'Doomwatch', 'Doomsday', the 'Doomsday Syndrome' and even the 'Great Doom Debate'. And of course there were counter attacks, telling us that the concern was excessive and distorted.

Biologists, and especially ecologists, were the central professional group featured as protagonists of environmental concern and indeed, to some extent, there was a divide between some life and physical scientists on the issue. When the 'Blueprint for Survival' provoked criticism from the Editor of Nature this, in turn, prompted a reply in the Times by some 180 prominent biologists who, whilst not supporting the excesses of the Blueprint, expressed their broad support for it.

Today we exist in a relative state of tranquility on the matter. Personally I hope it is tranquillity and not complacency, and certainly there is sufficient activity in the form of teaching, research and conference to justify this hope.

The environmental era stimulated a wide variety of educational developments. New subjects were introduced: environmental studies, environmental sciences, conservation studies, etc. Environmental topics were introduced into courses such as SCISP in science, and in geography and other subjects of the curriculum. This environmental education movement has been described as a bandwagon that has now turned into a hearse. This, to my mind, is an over-reaction and would be a tragedy if it were the case. An understanding of the many ways we affect and are affected by where we live and our use of natural resources, and a capacity for judicious decision-making concerned with the balance between environmental exploitation and human needs are, surely, necessary attributes for any citizen.

It seems to me that from this period we can learn, in particular, two lessons. The first can be inferred from what I have just outlined; that there is a need in education to consider what we mean by the quality of life. From the debates on environmental education we also learned that scientific enquiry is not a sufficient method by which to deal with social problems.

You probably know that ethylene is a constituent of car exhaust and have heard that it can damage plants that grow by the sides of roads. We can undertake an experiment to test the validity of this. Get three similar potted plants and enclose them in polythene bags. In one bag have normal air, in another car exhaust and in the third car exhaust together with a small amount of potassium permanganate which will remove ethylene. Keep the plants in a well-lighted room for two or three days and the one in the car exhaust (with ethylene) will lose its leaves.

Actually we would have to be somewhat more careful in our work than I have indicated but, clearly, what we have done is a scientific investigation testing a hypothesis and arriving at a reasonable conclusion: ethylene in car exhaust removes leaves from plants.

We can also undertake a parallel experiment; this time using three or four apples rather than exhaust gas. You will find that the plant in the bag containing apples without potassium permanganate loses it leaves. The others do not. The reason being that apples give off ethylene. This, again, is a scientific experiment.

However, if we now put the findings of the two experiments together we can have a social issue: What do we do about the pollutant ethylene? If you decide to ban cars producing ethylene in their exhaust should you also ban fruiterers along the streets? What would be the cost of taking ethylene from the exhaust or not having apples? What would people's reaction be to any decision we come to? (Incidentally, don't be worried; eating apples is not harmful!) Putting all the known facts and opinions we have together what is the best decision to make? Such now are the kinds of questions we would have to answer and, clearly, we are no longer dealing with a purely scientific matter.

The distinction between scientific inquiry and social decision-making prompts us to discuss a comparable distinction between what I will call Knowledge Studies and Issue Studies. The experiments with gas and apples are knowledge studies. When we put them together in a social context they become an issue study. The following table compares the characteristics of knowledge and issue studies.

Knowledge Studies	Issue Studies
Emphasis on knowing and understanding	Emphasis on problem solving and decision making
Unknowns tend to be known	Unknowns not necessarily known
Reflective	Motivation derived from social and/or individual action
Low moral load	High moral load
Precise terminology	Confused terminology
Unitary integration	Eclectic integration
Mainly invariable, possibly hierarchical, arrangement of concepts	Variable arrangement of concepts.

Comparison of the Characteristics of Knowledge Studies and Issue Studies

Let us consider another field to clarify the distinction further. Mendel's investigation of inheritance, referred to earlier, was a knowledge study. His aim was to know about heredity. It was a circumscribed package of knowledge he sought; a pattern of explanation. He also had a shrewd idea of what he was after and approached his work in a reflective manner. Because the topic of his study was clearly limited he was able, on the whole, to use words in an unambiguous manner. When he presented his ideas he was able to connect them together in an invariable logic: there were few ifs and buts in his explanations.

Even allowing for the inevitable vicissitudes of human endeavours this description, in essence, stands up, and particularly so in contrast to the way a problem in genetic counselling would be tackled. Here one is not seeking new knowledge but a re-assortment of knowledge to help solve a practical issue: for example, if the wife is rhesus negative should a married couple have children?

For different couples the answer would be different. They would be concerned with morality, not just correct explanations. How would they determine whether they have the right to possibly cause suffering to their own baby or pass on the possibility genetically to other generations? Religious views and cultural traditions may play a part. Even scientifically there would be more than one consideration. Besides the genetic probabilities of their children being affected there is also the relative efficacy of chemical treatment for an affected child to be considered. There are also broader social and economic considerations. It all adds up to a much more diffuse situation that is being dealt with than in the case of Mendel's study and even when the eclectic threads are pulled together, clear cut answers do not necessarily emerge. This is the nature of an issues study.

There is not an absolute distinction between knowledge and issue studies but given that - as the years of environmental concern have particularly highlighted - there is a need to link biological education (and that of other subjects) to the needs of the community as well as those of science, then it will serve a useful purpose by ensuring there is an appropriate blend of knowledge and issue studies in the curriculum. Issue studies link biology with the realities of peoples' lives.

I have attempted here to bring some matters to your attention which I feel are important in biological education and possibly have significance in the wider context of general education. I do not suggest what I have said is totally new. My discussion has not been finely honed. But, then, possibly it is better, in the circumstances, to aim at being roughly right rather than accurately wrong.

I have suggested that we should look for syntheses by which we can structure the content of courses in biological education and have stressed the importance of what I have termed the biosocial synthesis, which would involve close links between biological and social studies. A distinction is made between what I have called knowledge and issue studies,and I suggest that there should be an appropriate balance between them in the curriculum.

It is, argued that in biological education we should ask questions about the quality of life as well as the nature of life. As a component of general education biological education, in its broader and better sense, should intimately mesh the strands of scientific knowledge, a person's individual and community needs, and a consideration of moral values, into a firm educational fabric.

SECTION II

Reports from Different Countries

ASIA

BIOLOGICAL EDUCATION AND COMMUNITY DEVELOPMENT IN SOME ASIAN COUNTRIES

by A.N. Rao

The large percentage of world's population is distributed among the developing countries and people in many such countries live in poverty, hunger, squalour and various infectious diseases. The fruits of science and medicine appear to have hardly reached such peoples. The benefit of different biological training imparted in schools, colleges and universities have not made any significant impact on many human problems. The working document prepared and circulated by Professor Kelly has provided a wide basis for our discussion at this meeting. The points raised are applicable either in toto or partly to many of the developing countries. These include a) health and environmental problems, b) human biology and nutrition, c) education and community development, and d) the need for a proper balance between tertiary education, research and community development. The contents, materials and teaching methods are not excluded. The working document is thus a fairly complete and comprehensive one. It points out various problems experienced by many biologists in different countries, the only possible variations being the matter of degree, or the size of the particular problem may become variable, bigger or smaller, or certain problems may require more immediate attention in one country than in others and hopefully bringing suitable solutions in course of time.

In any society the role of a biologist is an important one. It is becoming more so whether it is the problem of conservation, environmental care, academic matters, nature education or management, food production or the improvement of human well-being. Many problems like these are not only complicated but also very closely interrelated. The biologists who work on such problems should have a sound and broad educational basis which will guide them in the right direction. This principle is equally applicable to all, whether he be a research biologist, a teacher or an administrator. They all have a key role to play in formulating the policies and projects for the improvement of mankind in future. How best we can train such a person suitable for the needs of the society, has to be reviewed from time to time.

Regarding the biological developments, I shall first discuss certain details about Singapore. According to the figures available for 1977, out of 2.2 million population, one in five of the population was in school of one kind or the other. The elementary biology is taught as a part of Science subject in primary 1-6 years. Biology is included as part of General Science in the first two years of secondary school, and during 3 and 4 years Biology or Human Biology is studied extending up to GCE 'O' level and later for another two years as GCE 'A' level subject. In the University, Biology is taught to undergraduates as two separate subjects Biology A and Biology B. There were about 6300 students in the University of Singapore last year and of these nearly 400 students studied biology. Regarding the course contents Biology A consists of courses that deal with the general organisation, classification, biogeography and ecology of plants and animals. Under Biology B the courses are concerned with Biochemistry, Plant, Animal and Human Physiology, Genetics and Microbiology. We believe that this system provides enough flexibility to include not only many aspects of plant and animal sciences but also to cover subjects like Anatomy, Biochemistry, Microbiology and Physiology. At the honours level the individual disciplines like Biochemistry, Botany, Microbiology, Physiology and Zoology can be studied as single subjects. If the students perform well at the honours year they can continue for their Masters and Ph. D. degree programmes. We find that the biology training given to our graduates is adequate and prepares them well to discharge their duties satisfactorily for the welfare and the improvement of communities in Singapore. This applies equally well whether they become teachers, administrators, scientists or take up other professions.

The biology courses at Nanyang University in Singapore, Universities in Malaysia at Kuala Lumpur and Penang have comparable teaching curricula. The duration of training is also similar.

Now I shall discuss certain details about other countries in Asia. Malaysia has a population of about 12 million and there are 4 universities imparting Biology education including Agriculture and Forestry. The biological subjects in these universities are very varied both in scope and contents to meet the different requirements of the country.

In Thailand, for a population of 44 million there are 23 universities (13 state and 10 private) and in 1978 there were about 3000 biology undergratuates and 250 postgraduates.

In Indonesia, with a population of 120 million, there are about 30 state universities, classified as 5 leading, 9 intermediate and 16 less developed universities with a total Biology student population of about 3500 students.

In the Philippines there are many state colleges and universities. They accomodate a very small proportion of college population. Admission is not controlled and therefore the standards achieved are very variable. At the University of the Philippines, Los Baños, there were 5586 students enrolled last year and of these only 122 studied undergraduate biology. There were approximately 45 undergraduate students to every graduate student.

In Sri Lanka, Biology is taught in the high schools both at 'O' and 'A' levels. Not more than 15 % of those who qualify for 'O' level proceed for further studies. Of the 60,000 students that sit for 'A' level examinations about 27,000 (less than 50 %) qualify for university admission but only 5000 could be accommodated in the universities. The degree structure is based on 3, 2, 2 subjects system.

Bangladesh has a population of 80 million people in an area of

55,000 sq miles. Urbanization is in its early stages, less than 10% of the people live in settlements of more than 5,000 but it has started to increase at an alarming rate. The details of Biology curriculum taught in Bangladesh are similar to those in India and Pakistan. Shortage of teachers, lack of equipment and facilities are common problems in that country. There are five universities in the country providing general education. Postgraduate programs in Biology leading to M.Sc. and Ph.D. degrees have yet to be properly developed.

In India many fundamental changes are made to the curricula to bring the biology syllabi up-to-date, in different universities. There are about 60 universities at present. A co-ordinated programme to unify the principles of biology at different levels of organisation, both of plants and animals has been prepared and adopted. Such a programme is followed both at undergraduate and graduate levels. Compulsory courses in Biochemistry, Biostatistics, Biophysics are introduced and the Masters degree programmes in Life Sciences have been started instead or in addition to the regular programmes in Botany or Zoology. In the new universities Biology departments or divisions have been started. The College Science Improvement Programme is launched to upgrade the biology teaching in undergraduate institutions through improving the methods of instruction, curricula, facilities, equipment, library and teaching materials. The results achieved so far indicate that the programme has made a real impact both on the teaching curriculum and quality of teaching.

In Papua New Guinea, biological knowledge and its application is gaining importance in various community developments. The Man and Biosphere (MAB) National Committee was established in 1977 which is working on three major projects. The projects comprise many component basic biological studies that vary greatly in intensity and scope. These are a) mapping of vegetational types, b) inventory of commercial timber species, c) natural regeneration of forests, d) ethnobiological survey, e) social impact of forest industry on village communities, f) studies of local populations including human nutrition, parasites, immunological status and g) interrelationships between environmental conditions and life-style on the one hand and human health and well-being on the other. The working group includes natural and social scientists from the two universities and representatives of governmental and non-governmental institutions.

After outlining some of the main points of biological education in different Asian countries, we may now consider some of the biological problems characteristic of developing countries including inadequate nutrition, undernutrition and malnutrition with the deficiency diseases that follow, various diseases like Malaria, Filaria, Leprosy, Tuberculosis, Cholera, and other liver diseases. The medical doctors are produced in urban areas to serve the urban populations with an imbalance ratio between urban and rural doctors, 1:500 in urban areas, 1:1000 or more in rural areas. To meet shortages of doctors, especially at the time of Cholera epidemics, crash training programmes of 1-2 weeks were carried out in Bangladesh. The service rendered by such paramedical personnel saved many from death. There are more infectious diseases in developing countries than in developed ones. There is a definite correlation between the development of a country and the disappearance of infectious diseases and the appearance of heart, vascular, metabolic diseases and cancer. Many of these can be controlled by improved sanitation, supply of good, clean water and providing proper foods. The biological knowledge and its application in food industry is important, especially in nutrition, food processing, preservation and others. Such solutions are closely tied up with socio-economic and political problems of a given country. Elementary and fundamental biological knowledge spread by mass media may bring beneficial results. Leprosy and Filaria are mostly localised in India, Indonesia, and they can be eradicated by determined efforts as has been done in case of Small Pox.

The food and population problems are more involved including agriculture, storage of grains, distribution, good management etc., and for the brevity of time the details cannot be discussed here. But the students should be made aware of these topics since the food problem is one of the fundamental aspects of community development.

Bioscience education can solve many problems created by modernization that makes heavy demands on environment. The conflicts between development and environment can be solved. The educational aspects dealing with the improvement of environment, have to be stressed,and ecological, environmental studies should complement each other. This is equally applicable to both rural and urban environments. Introduction of ecological studies into Bioscience education is important and students while studying the environment will get close to nature and appreciate the problems involved that upset the balance in natural surroundings. Subjects of Ecology, conservation, pollution and renewable energy sources have great socio-economic implications and these should be introduced in curriculum development. This is more urgent in the developing countries, where forests are cleared to make way for agriculture, human habitation, industries and other purposes. Such activities bring a train of undesirable changes in the environment to the discomfort of so many. The conservation of nature and genetic resources are equally important for the improvement and benefit of mankind in future. This aspect of the problem is more relevant to tropical countries, their fauna, flora and other resources. In this connection courses on 'Man-environment interaction' have to be effectively taught at different levels both for Science and non-Science students.

One of the critical issues involved in Biology teaching concerns the text-books and teaching aids. Every region or country will have its own biological problems and variations. The text-books produced elsewhere may not illustrate or be applicable to the local conditions. Text-books using tropical plants or animals to illustrate different biological principles are very few. The local diversity of climate, ecology, human population, fauna, flora should be properly illustrated in the text-books used. Indigenous talents have to be developed encouraging the teachers to write books that are relevant to their country or region, and the money should be made available for the publication of such text-books by local governments. Where such help is not forthcoming the international organisations should come to their aid.

Some of the questions posed in the working document are "Can more research and development work in biological education related to community development be established in universities? What are the essential features of such work?" and the others on page 5, para 6.

I shall deal with these questions very briefly. As the following two tables would show there are quite a number of journals published in the developing countries and nearly 64 % of these are in Biosciences. But Bioscience research has not yet developed to any satisfactory level in majority of these countries. Apart from the number of papers published the number of research workers and the money available for research is small in the developing countries. The biologists in developing countries aspire and attempt to publish their papers in international journals as against the local journals to obtain recognition in their own fields. Very frequently the research carried out may have little or no relevance to local problems. Applied research, for the development of the community and its needs, has hardly begun in the developing countries. Greater efforts should be made to investigate local problems and their solutions. If the effort is rewarded by the use of the community, such a research is very worthwhile even though it may not receive very high international recognition by the scientific bodies or learned societies. Thus the greater part of research should be utility-oriented. Research problems are many whether they are of food, agriculture, health,

TABLE I: Number of Scientific Journals published from Developing Countries in Asia and Africa

Country	Scientific Journals	Bioscience Journals
Egypt	1	1
China	8	7
India	86	52
Iran	1	1
Kuwait	1	1
Lebanon	2	1
Malaysia	1	1
Pakistan	2	2
Singapore	3	-
Taiwan	8	6
Thailand	1	1
Total	114	73

Ulrich International Periodicals Directory, 16th edn., 1976.

TABLE II: Publications in Biosciences in Developing Countries in Asia and Africa

Country	Number of publications	Country	Number of publications
Bangladesh	7	Libya	2
Botswana	1	Malaysia	31
Burma	4	Nigeria	131
China	8	Pakistan	22
Egypt	96	Philippines	2
Ethiopia	2	Rhodesia *	25
Ghana	5	Saudi Arabia	33
Hong Kong	39	Singapore	35
India	1260	Sri Lanka	14
Indonesia	3	Sudan	5
Iran	37	Syria	1
Iraq	25	Taiwan	10
Kenya	26	Thailand	25
Korea	1	Turkey	33
Kuwait	9	Uganda	1
Lebanon	5	Zambia	8
Asian Region	1574	Total for Asian and African Regions	1876
African Region	302		

Current Contents - 1977.

* Now Zimbabwe

communicable diseases and others. People in authority and government should encourage the local scientists and develop a good line of communication. There is no doubt reason why such research connot be programme to yield beneficial results.

The contribution of Biosciences plays a significant role in improving the quality of life in developing countries. In Indonesia a sanskrit word 'TriDharma' (Three righteous deeds) is used in the mission of higher learning as education, research and public service. All these three are equally important in promoting biological education towards community development. In the above account I have presented certain data gathered from various Asian countries and these I hope, would serve as useful background information. We should remember that in these countries the problems are varied in size and scope requiring both quantitative and qualitative solutions and in certain cases very urgently. Our present meeting is organized to identify various but specific objectives. It is hoped that to define and develop the proposals or objectives the points made in this paper may prove to be very helpful. Biology for the sake of Biology and Knowledge for the sake of Knowledge are luxuries that few can afford, even in developed societies. The knowledge has to be put into good use that would help the community in many ways.

BRAZIL

BIOLOGICAL TEACHING IN BRAZIL AND COMMUNITY DEVELOPMENT

by Myriam Krasilchik

Education in Brazil is divided into three levels: the Fundamental school, compulsory by law, includes the first eight years of schooling and is destined for students between 7 and 14 years of age. The Middle School lasts for three years and includes students from 15 to 17 years of age. At the University, the students already choose specific, professionalizing courses. Only in some colleges there is a basic or common one year course on general education.

The Secondary or Middle School has always been a formal, traditional course, having the University as its aim. In 1971, a new educational law, destined to change the situation, determined that the Middle School should have as its objectives not only the complete education of the youths but also their preparation for work.

To achieve those aims, the middle school curriculum is formed by a group of disciplines of general formation and a set of activities for special formation to be established according to the demands of the different regions of the country.

For administrative purposes, teaching is organized into systems: the federal system and the state systems. The directive agency of the federal system is the Federal Council of Education that in certain cases can decide about the state systems that are under the authority of the State Councils of Education and the State Education Offices. There is a net of public schools and a group of private ones, under federal jurisdiction, maintained by religious institutions, minority groups or profit aiming organizations.

The curriculum as to general education in the fundamental and middle schools and as to the professionalization (part of the middle school special preparation) is fixed by the Federal Council of Education but the state systems can establish disciplines in the diversified part with which the schools complete their curricula.

The Present Situation of Biology Teaching

Within the range of disciplines in the Fundamental School there is a place for the Physical and Biological Sciences, including topics in Physics, Chemistry and Biology that are taught by the same teacher. The aspects that are usually taught in that course of "Combined Science" include elementary notions of morphology and systematics, elements of human physiology and anatomy, elements of cytology and genetics and some aspects of hygiene and nutrition.

In the Middle School, the recommended program for the first year common to all the areas includes aspects of General Ecology. It is admitted that besides touching extremely relevant aspects, it is a structuring concept that leads to other discipline areas that appear in the syllabuses in the subsequent years. In a survey made among the Biology teachers in the State of Sao Paulo, the following topics were considered, in a decreasing order of priority: 1) Ecology; 2) Physiology; 3) Genetics; 4) Morphology (Krasilchik, 1979).

A project of preparation of material to help the teacher is being developed by the Science Center of Sao Paulo (Cecisp). The conception of the material is based on the idea that the fundamental and most difficult thing to achieve in relation to innovation, is to change the teachers' methodology, really engaging the student in activities that demand a logical and critical reasoning enabling him to work with the community. A series of activities that demand little equipment but involve an active participation from the students are being tested. The aim is to transform bookish teaching into an activity directed to action.

There are also some attempts to develop projects in rural communities or outskirts of towns (the Science Education Center in Bahia, with support of the Rockfeller Foundation); the anteproject of the State of Pernambuco Office of Education; Environmental Education Project (at Ceilandia in Brasilia). The majority is still at the initial stage and they are limited to some regions in the country.

Future Perspectives

The relationship Biology Education-Community is without any doubt a two-directional process: school-community and community-school.

Therefore it seems more reasonable to use the school, an institution that already has an established role within the community for function as a basis for a series of programs that will make Biological Education more socially relevant. It is true that, in general, the school is a conservative institution and it may be difficult to try to produce a renewing action inside of it. However it is possible to use that potentiality to cause changes whose limits can be foreseen.

Besides that, it seems to us to be essential to possess the knowledge of factors that interfere in the curriculum and to use them to obtain short term positive results, in a realistic and viable manner. If it was verified that it is easier to obtain the transformation of the taught contents, a first goal for the work to be done should be the introduction of topics like the ones listed in the Report of the Meeting - Biological Education for Community Development, 1978 - in the Biology program.

As it was proved that the text-book plays such an important role in determining the real curriculum, it seems to us that in Brazil, an urgent step towards relating Biological Education to Community Needs should be to elaborate a didactic book to promote not only the content transformation but also the methodological transformation. We have observed that teachers, upon receiving teacher's guides with activities instructions, start to ask for material prepared in advance for the student under the form of a text-book.

This fact is made clear when we verify that the teachers that work with curricular projects demand methodological instructions what does not happen to those who do not have experience with projects. Consequently, it seems to us that the most practical course of action to take should be to elaborate a modular project formed by the student text-book and the teacher's book following the same orientation we had in the 1960s.

With the twenty years of experience, it will be possible to make realistic projects, that correspond to the students' necessities, causing a true change and making the school more sensible to the community needs.

Some special cares have to be taken in order to better the project's chances of success:

a) People with different origins and backgrounds should be included in the writing team, including sanitarians, agricultural engineers, nutritionists, biologists, biology teachers actually teaching, and education specialists.

b) University Institutions, dedicated to teachers' training and other university units should be involved during the project's developing period. Research referring to several types of training periods that leads the student-teacher beyond the classroom to an incursion in the community life is urgently needed.

During the future teachers' training period, a strategy to make them aware of their changing role will be to add community services to their training activities. That will put them into contact with the real problems faced by the students they are working with.

Summarizing, we believe that science education has now reached its baroque phase: either we go back to classical simplicity and thus to a renaissance in the field or we run the risk to suffer a declining phase.

Bibliography

KRASILCHIK, M. - Uma experiência na Renovação do Ensino de Ciências - Anais do Simposio sobre o Ensino de Biologia, Fisica, Matemática e Quimica - Publicacão ACIESP, no. 11, 1978.

KRASILCHIK, M. - Teaching biology in Brazil - Journal of Biological Education - it will appear in Vol. 3 (4) or Vol. 4 (1).

LAWLOR, Hugh Antony - Participants and Influences in the Process of Educational Innovation in Brazil, with reference to the reform of science education at school level - Unplished PhD. Thesis, The University of Hull, 1978.

MAYBURY, R.H. - Technical Assistance and Innovation in Science Education - Science Education - John Wiley & Sons, N.Y., 1975.

RAW, I. - An effort to improve Science Education in Brazil, - Mimeographed, MIT - Boston, USA, 1970.

CANADA

BIOLOGICAL EDUCATION FOR COMMUNITY DEVELOPMENT IN CANADA, LOCAL PROGRAMMES, ACTIVE AND POTENTIAL

by F.A. Gornall

It is not possible to discuss biological education for all of Canada in a short essay because there are at least thirteen autonomous educational systems within the country. Each province has complete control over its own educational programmes. These programmes vary considerably in biology as well as other subjects. I shall comment therefore only on biological education in the secondary schools and on teacher education institutions in the province of British Columbia.

For many years B.C. has had a province-wide curriculum prescribed for each discipline by the Ministry of Education. Biology was interspersed as units or chapters from traditional texts in courses of junior secondary science. Thus biology sections were usually taught by a biology teacher. In senior secondary schools two courses of biology were prescribed, one for all science students and a second leading into biology at the tertiary level. Examinations administered and graded by the Ministry became the great equalizer.

A traditional descriptive course of dictionary biology was shelved to make way for the enquiry approach exemplified by the BSCS yellow version. Unfortunately the enquiry approach was rarely allowed to happen and descriptive biology continued mainly as a reading exercise from the Yellow text. Now the Yellow Version and its enquiry approach has been superseded by "modern" traditional biology texts. Recently provision has been made by the Ministry to allow individual schools or teachers to submit locally derived curricula for Ministry approval. Although the number of classrooms involved is small as yet, it is interesting to note that most proposals for alternate programmes are based on community needs or demands. These propose or embrace programmes in forestry, fisheries, marine biology, agriculture or general environmental programs. They are usually based around and supported by a local industry but are put together as courses by individuals or small groups of teachers.

Much more common though is to include units of work as a portion of a descriptive biology course. Several weeks may be spent on an immersion type programme of local interest. i.e. a comparative study of salmon development in a stream bed with salmon development in the classroom. The activity had the full support and cooperation of the Fisheries Department. Or a forestry program involving the collection of conifer seeds from "supertrees", genetics and sylviculture in local nurseries and in the classroom again with the co-operation of the forest nurseries. Outdoor environmental education has mushroomed in the last ten years. Many schools and school districts have their own sites and are developing field study centres. Co-ordinators and consultants have been appointed by school boards to assist teachers with these studies. Unfortunately this type of programme has not caught on as well in the secondary schools as in the primary schools. The stumbler seems to be rigid secondary school time-tables. Enterprising teachers have sidestepped this block in a number of ingenius ways.

One enthusiastic teacher has acquired a suitable vessel and organized a marine biology program. The vessel is available to school groups who wish to pursue a study of marine biology. Trips last 4-5 days and include far more than the text book definition of marine biology. Effects and degree of pollution are high on the list of investigations.

The University of British Columbia owns a 12,500 acre research forest about 50 miles from the main campus, operated by the Faculty of Forestry who share its facilities with other faculties and with elementary and secondary schools. In addition to forestry operations and forestry research programmes there is a Monday to Friday elementary school resident programme. In addition teachers may arrange their own special programmes. Much of the property is open to the public on week-ends, primarily for recreation but with an increasing emphasis on public education.

The Canadian Forestry Association in co-operation with the forest industry and our Faculty of Education organizes courses of 2-3 days duration on various aspects of forestry. Instruction is given on site at sawmills, pulp mills, logging operations or nurseries.

Commercial lumber companies provide four tours a year for groups of 24 students, 12 forestry students and 12 education students. The tours are from four to seven days and include visits to most of the operations related to forest industry.

The Vancouver Public Aquarium provides laboratory facilities and operates daily programmes for biology students at secondary and tertiary level. Teachers and student teachers may also use this lab for their individual studies or to conduct their own programmes. The Aquarium also provides an instructional and training programme for about 300 docents who in turn interpret displays and events to the public.

VEEP - Vancouver Environmental Education Programme - based in the Faculty of Education at the University of British Columbia provides an opportunity for talented teachers to share their ideas and classroom experiences with others. VEEP is supported mainly by grant money. Teachers are hired or provided with release time to develop and report on their own innovative projects. Work is published and the attractive booklets are distributed at cost. A current contract with the Canadian Cancer Society will involve teachers in the development and preparation of a community cancer education programme.

CARIBBEAN

BIOLOGICAL EDUCATION AND COMMUNITY DEVELOPMENT - A CARIBBEAN PERSPECTIVE

by W.K. King

Introduction

The Caribbean region, as this paper conceives of it, stretches from Belize in Central America to Guyana in South America. These relatively small islands and the two mainland countries bear some striking resemblances socially, culturally, politically, economically and educationally. Yet, in many respects, they are completely different.

The fifteen countries are at varying levels along the development continuum. Some are fully independent, some are only internally self-governing, with Britain maintaning responsibility for defence and external affairs, a few are still colonies of the United Kingdom.

Many of the countries have no natural resources and depend almost entirely on agriculture and tourism. This makes survival rather difficult and development, if any, very slow.

As such then it is not easy to meaningfully cover all aspects of the topic in detail. This paper will of necessity be a brief survey of the ways in which biological education can aid and is aiding in community development.

The Role of Science Education in Development

It is perhaps important to look at the ways in which science education should feature in developing countries and then zero in on the whole idea of biological education and community development.

Professor Fletcher Watson in a recent paper puts science education in the driver's seat, so to speak, in developing science for survival: 'At least part of the responsibility for developing the needed competencies for making wise decisions resides in the area of science education. We must help students to learn to appraise availiable information and rationally analyse the long-term environmental and social consequences of alternate possibilities for action or restraint'. (1)

The author of this paper in a recent article delineated three distinct areas where science education is essential in a developing country. These are (a) science as a vehicle for learning: 'The method of enquiry, methodical imagination, disciplined speculation are all essential to any learning situation' (2); (b) science as it informs technology: 'It seems that if technology is not science-based it cannot progress (3); (c) science for scientific literacy: 'What we need in developing countries are people who through training have acquired a certain attitude and concern, especially in science of the environment'. (4)

To substantiate the views of Professor Watson and the author one sees today all countries - developed, developing, underdeveloped, making deliberate efforts to fashion an education in science that is consonant with their needs and aspirations. At no other time in the history of mankind has science education been called upon to deliver so much. Some educators are understandably alarmed at the increasing role they are called upon to play, some are bewildered.

Biology, in that it touches man and life most intimately, is possibly the area most in vogue. Biological educators in the face of the thrust towards environmental concerns find themsevles overwhelmed by demands from all quarters. The overall effect is a frantic search for a better life for mankind and his offspring.

Here are a few of the trends in biological education that are evident in the Caribbean region (1) towards more and better food sources (2) towards 'children by choice and not by chance' (3) towards eradication of disease (4) towards an aesthetic education in biology (5) towards an understanding of racial peculiarities (6) towards an understanding of mental illness.

1. Towards more and better food sources. There was a time when the bulk of the food was imported. In some countries there is still some trace of this practice. However, there are many areas of research in which Caribbean people are searching for more and better varieties. By and large these varieties are expected to give (a) more yield (b) greater resistance to known diseases. The cases of the Barbadian sugar cane and the arrowroot from St. Vincent are just two of a very long list.

Biological education is also laying great stress upon replacing imported foods with local substitutes. It is common knowledge that Caribbean peoples have acquired certain tastes in food at the exclusion of what is local, cheap and mostly readily available. The need then to educate people about the nutritional values of their crops, and their comparability with temperate ones is most essential. Take for example the common fallacy that blackcurrant juice must be the sole source of vitamin C. The Caribbean region is blessed with several fruits which are good as, if not better, sources of this important food:

A comprehensive guide to common Caribbean foods has just been released. This document is being used at all levels of education in Biology throughout the region.

Allied to the search for more and better food crops is the establishment of agro-based industries. These centres, for example CARARI in Trinidad and Agrolab in St. Vincent, are looking for modern and effective methods of food preservation. Preservation will allow the region to handle gluts and earn valuable foreign exchange from export.

There is also a new effort in the area of fisheries and animal farming. Many countries have embarked on programmes of studying life cycles of marine life to determine the best time to catch them. As a result there are 'closed-seasons' on most economically important marine life. Here again such ideas are filtering down into the biology curriculum of learning institutions.

It is understandable that the old eating habits of people do not change overnight. Biological education seeks to develop the community not only in providing knowledge and skill, but also in developing attitudes. It is only when parents change their eating habits and communicate the need for change to their children, that any impact would be made in this area. Knowledge that our ground provisions are in no-way nutritionally inferior to wheat will continue to be inert until we put that information into action.

2. Towards 'children by choice and not by chance'. Within the last two decades there has been a shift towards family planning. The region has realized that with little natural resources and the pressures of international inflation, there is a need to slow population growth.

Each territory has some form of family planning programme and, according to statistics they seem to be effective. Here again, however, there is a big problem of attitudes. Many persons believe that family planning methods encourage promiscuity. They are afraid that the moral fabric of society will fall apart.

Sex education on a time table is still likely to cause raised eye-brows. Great care is taken not to give children any idea about their own sexuality. Apparently, they find out anyway. Unless it is understood that the time is right for teaching about sex in schools - using qualified, suitable teachers, of course, the children are likely to get sordid, warped impressions on the topic from their peer groups and the media.

3. Towards eradication of disease. For a community to develop it must formulate strategies to eliminate disease. In the Caribbean there has been a trend towards more effective methods of garbage disposal, which should go a long way towards preventing disease. Now more than ever people are most conscious of the dangers of harbouring flies, cockroaches, rats, mosquitoes.

Disease cure is also under serious research. Using knowledge and skills from Biology there is a thriving set of programmes to find out what remedies, if any, are present in local herbs and plants. Some startling results are coming to light e.g. two ophthalmic surgeons have found marijuana to be helpful in treating some ailments.

All these findings are being put into biology curricula at all levels.

A whole branch of the study of disease, namely deficiency diseases, is taking prominence too. This is co-ordinated with the emphasis on foods and their nutritional values. There is hope that old taboos and superstitions would evaporate eventually.

4. Towards an aesthetic education in Biology. So often Biology is taught as a subject devoid of any aesthetic concern. There is, however, inherent beauty in nature and biological education should tap such beauty. Especially in a region so full of all types of beauty it is an error to miss this opportunity.

There was a time when biology teachers insisted on drawing flowers etc. on the chalk board when a few metres away there were infinitely more beautiful specimens. This habit is definitely on the decline. It is important that one realizes that this is all part of the general move towards emphasis on study of the environment. However, it is possible to study the environment without pin pointing the beauty in nature. There is room for still more effort in this trend.

5. Towards an understanding of racial peculiarities. This trend is probably least pronounced, although it is so important. One of the most vital aspects of education is that a person learns about himself or herself. There is need for biological education to emphasize why Caribbean peoples are mainly black, why certain diseases are peculiar to black people, why there is nothing inherently inferior in being black.

To the author this is essential to any liberation and development of person and community. Many persons, however, shy away from this topic because it smacks of racism. This stand is untenable. Racism involves much more than knowing what is the use of skin pigmentation. If any thing, ignorance allows one to harbour doubts, and maybe resentment.

6. Towards an understanding of mental illness. A person who is mentally ill acquires a stigma which lasts for years. The study of the nervous system and how it works has opened up new vistas in treatment, care and rehabilitation of those afflicted.

Another aspect of this trend is the new feeling towards the mentally retarded. Society is now developing ways and means of using whatever skills they have to develop the community. The advances in this field could be attributed to biological education, especially in the area of human biology.

Some Limitations on Biological Education

It is important that one realizes that biological education is not doing all that is possible to help in community development. The two main reasons are (1) that external examination syllabuses act as constraints (2) a somewhat strange notion of what community development entails.

To concentrate too much on the aesthetic nature of biology, for example, is to run the risk of being accused of 'not following the syllabus'. This is a grievous accusation, especially when one realizes that an examination pass is a passport to employment, further studies and even social mobility. The author sees two main ways of overcoming this problem. The first is to use informal, community-based education to supplement the formal school-based instruction. In Caribbean countries where television is a way of life there are frequent programmes on dental hygiene, birth control, disease prevention und cure. This method is, however, in its infancy. The second way is to develop indigenous programmes that place emphasis on local conditions and needs. There are obviously many other ways but these two seem relatively easy to implement.

To many people community development means building community centres, self-help housing development, and activities concerned with outward and visible signs of progress. To the author there seems no other way than to start by educating the individual about himself, his environment, his country, his world. Visible trappings although necessary are but the tip of the iceberg. Development must be all round and deep.

Various Ministries of Government handle community development in various ways, but generally speaking it is included in the endeavours of all countries. What is questionable is some of the approaches. True, there is scarcity of money, but community development projects should be self-generating. This last point further emphasis the need to use biological education as the framework on which to build. The person who keeps a dirty backyard and encourages flies which his neighbour fights valiantly against, will soon receive sanctions to help him change his ways. The biological concept of 'community' is indeed useful to explain to him the errors of his ways.

Role of Biological Education in Community Development

It is interesting to note that the three distinct areas which King (1978) delineated as essential roles for science education are present in the six trends listed above. Let us look at each role in turn.

Biology, with its heavy emphasis on observing, describing, reporting, measuring, by its method of enquiry is essential for teaching people how to learn. In developing countries like the Caribbean, where lack of resources forces certain constraints on formal and informal education, it is necessary to train people for self-initiated and self-directed learning. One cannot claim that Biology is unique among the sciences in producing such skills. However, because in Biology one works with living things with which one can identify, there is an instrinsic motivation that it produces. This is one reason why so many more students do the subject. Another reason, of course, is that it is the 'softest' of the sciences because it deals with mostly visible, concrete examples. The author believes that this appearance of simplicity is very deceptive.

Then take the point of biology informing technology. Agriculture and fishery spring readily to mind. But there are others, such as forestry and pharmacy. In the Caribbean there are several subject areas which carry names like Health Education, Family Life Education, Home and Family Life which, if one delves deeply enough, are nothing but biology and its social applications. These represent offshoots of biological education, made by man to allow knowledge to be handled in more manageable packages.

The notion of scientific literacy is well aired in developing countries. As man progresses he comes to realize that without proper shepherding of his skills and knowledge he may yet produce his own destruction. A citizenry which has a working knowledge of aspects of the environment as they affect life, is the aim of all our countries. Even more important is that the policy-makers must be sufficiently trained to make proper decisions on reasoned judgements: "My central concern is not for those who choose careers in science, but for the large 'public of the future' who as planners, bankers, lawyers, legislators and citizens decide individually and collectively what applications will be made of the possibilities revealed by science-technology." (5) Biological education is being called upon to eradicate the superstitions of Caribbean people and to replace these with proper reasoning based on cause and effect. As mentioned before, it seems that this scientific literacy is the first step in a truly developed community.

Summary

Biological education aids in community development not only in providing knowledge and skills but also in the fostering of proper attitudes. As such then the task of biological educators is not an easy one. Yet their contribution is so essential that it must be carried out.

The Caribbean region as it stands at the cross-roads of social, cultural, political, economic development needs the method of enquiry and the disciplined approach to problems that biological education offers, to help solve many of the problems that development brings. An educated community is the first essential step to building a developed community. Science education generally, and biological education in particular, is a powerful catalyst in such development.

Bibliography

WATSON, F.G., Science Education for Survival. Paper delivered at International Symposium on World Trends in Science Education, Halifax, Canada, August 1979.

KING, W., The Role of Science Education in a Developing Country, The Journal of The Ghana Association of Science Teachers, Volume 17, No. 2, January 1978.

KING, W., op. cit.

KING, W., op. cit.

WATSON, F.G., op. cit.

ISRAEL

BIOLOGICAL EDUCATION AND COMMUNITY DEVELOPMENT IN ISRAEL

by A. Blum

The theme of Community Development appears in various areas of the school curriculum, not only in biological education. But biology and related subjects are expected to contribute specifically and do so probably more than the more descriptive and static subject Civics.

There is a tendency to start at an early age with the close environment of the child - the home and neighborhood, and only later to treat developmental issues of larger communities - city, country and the whole human biosystem. This can be illustrated by developments in Israel.

Fig. 1 shows the situation diagrammatically.

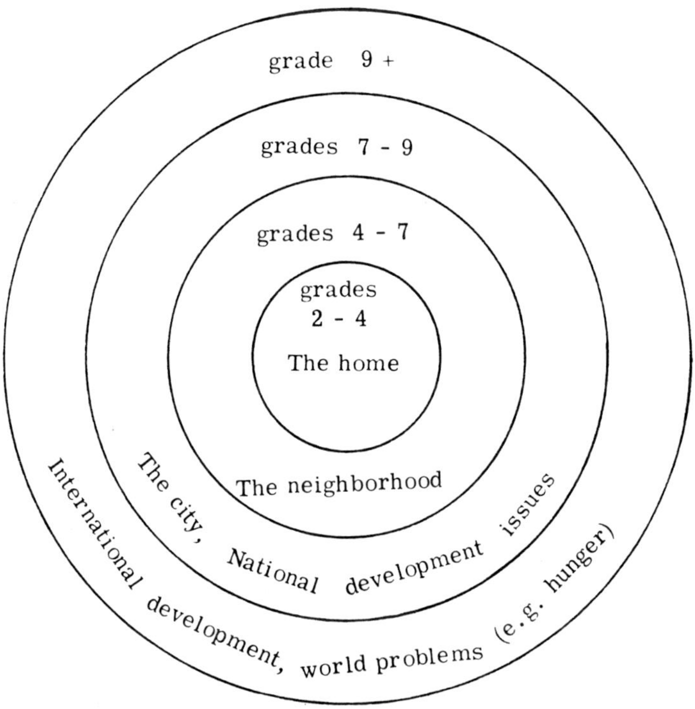

In grades 2-4 (ages 8-10) the topics The Home and The Neighborhood are taught in the framework of social studies. In Environmental and Agricultural studies (a strongly biology based subject), 5th and 6th grade pupils (aged 11-12) learn how to beautify their school by planning and planting an ever-flowering corner of annual plants and caring for them. Also they take potted plants home and discuss the beautification of the neighborbood in which they live and learn.

The city and how it developed is treated in geography, at the Junior Secondary level (ages 13-15). The historical development of various communities is discussed, separately, at about the same time, in history lessons.

Biology (compulsory) and Environmental Studies (recommended, but not compulsory) deal with community development issues mainly at the national level, and in some cases international development issues are investigated. This is done mainly in 9th grade - at present the last year of compulsory education - when students are about 15 years old. At that stage pupils are supposed to have studied animal and plant ecology and some human ecology. The biology curriculum recommends that at least one instructional unit each (out of about six which can usually be managed during the school year) should be in each of the three main themes: "Man's Interference with Nature", "Man's Peculiarity in the Animal World" and "Unity in the Living World". The first two themes are clearly related to community development issues (with a reminder of the danger of upsetting the balance of nature by unwise developments).

The topics suggested by the 9th grade biology curriculum and relevant to community development are the following, and they are not exclusive:

a) Reciprocal influences between man and landscape.

b) Water pollution and ways to conserve their quality. ("Simulation games may be used to enhance pupil's awareness of environmental pollution. Groups on issues like assessment of the economic and social implications of the preventive means discussed").

c) Use of pesticides, their advantages and dangers - "It is advisable to discuss alternative solutions to chemical pesticides ... The need for a balanced solution and to abstain from one-sided approaches should be stressed".

d) The problem of the growing world population and its food supply needs - including agrotechnical, industrial and futuristic improvements, and the need for international cooperation. A simulation game is used.

e) Efficient use and recycling of material resources.

f) Industrial pollution (e.g. air, noise) and its effect on the environment - ("the problems arising from urbanisation and industrial development and ways how to improve them").

The majority of these topics were developed by the Agriculture as Environmental Science (AES) project, but were included as options also in the biology curriculum. More detailed descriptions of these units can be found in the brochure "AES". (Some English translations, e.g. The World Hunger Problem, are available for study).

In the upper secondary grades developmental topics are less common, because the prevalent text is an adaptation of the Yellow version of BSCS. One of the modules originally developed in Israel and recommended to biology teachers in the upper secondary schools is on plant hormones and their central role in agricultural development.

Students majoring in Biology, at their final secondary school examinations, can offer a written investigation on any topic, including developmental issues, instead ot the set examination.

At present, a new integrated school subject - Production Ecology - is introduced as a central programme in Agricultural and Rural-Regional Secondary Schools, leading up to matriculation. This subject integrates aspects of Biology taught in Secondary Schools, plant and animal growing and breeding, as well as Basic Economics. It is taught in 10th-12th grades, approximately 6 periods a week, which is a very high concentration for Israeli schools, in which most subjects are taught only 2-4 periods a week.

In Kibbutz schools less emphasis is put on agriculture, but the social aspects of biology are stressed.

With all these promising curriculum developments, practice looks often more traditional and dull. The main obstacles to a much wider implementation of these curricula in school are teacher's reluctance to get involved in unfamiliar problems, and to some extent, a lack of their public demand for more school involvement in community development.

At the university level the situation seems to be similar to that in other countries. Subject matter specialists are reluctant to combine "pure biology" with community development, a topic which is usually looked upon as a social science speciality. Therefore usually a team approach, at a highly advanced level (mainly in research at the post-graduate level) is recommended. New study combinations, like Environmental Science, do exist, but many university teachers view them with suspicion.

JAPAN

COMMUNITY DEVELOPMENTS AND BIOLOGICAL EDUCATION - A CASE STUDY OF SOCIOBIOLOGICAL OR BIOLOGICAL EDUCATION

by K. Imahori

Historical

Biological education and community development in Japan progressed in superficially similar ways, although the motives of change differ somewhat. During beginning of the 1960's the innovative work of BSCS gave a strong shock to many Japanese biologists and biology teachers, especially when the blue version of BSCS was introduced. Biology text-books of the upper secondary school level above all, were affected and educational confusion resulted. That is to say, many Japanese biology teachers misunderstood innovative idea of biological education, and as a result, high level knowledge acquired a priority in biology education. Many teachers were convinced that to teach biology as a science should be the aim of science education for secondary level classes. Under influence of the tendency, living organisms were ignored in biological education. Not only pupils but also biology teachers became ignorant of the names and structure of organisms though they studied living phenomena in molecular level in detail. The worst was that the biased biology education fitted to peculiar Japanese community which is oriented higher academic career. Extraordinary keen entrance examination for colleges and universities demands scientific knowledge, and many teachers of the upper secondary school level are diligent in preparation to make their pupils successful in entrance examination. Understanding and remembrance of biological knowledge are respected in classes whereas laboratory works are held lightly.

Criticisms and introspection have been made rather recently in Japan, of course. Environmental and energy crisis of the world have influenced on those introspection, too. Some biologists gave warning against the tendency of biology education. In the some circumstances new idea was proposed that the "biology" should be taught in connection with the human itself or human society. The idea was adopted first by the tertiary school level, especially in liberal education. Educational criteria of each school level are controled more or less by the Ministry of Education which announced to amend the criteria of liberal education in September 1970. The amendment had made it possible to provide interdisciplinary subjects. Biological education of the lower level is also changing slowly which will be explained later.

I. Interdisciplinary Subjects and Biology

A marvellous progress of scientific researches brought about subdivisions of biological sciences such as cell biology, cytogenetics, population genetics, immunogenetics, biosystematics and so on. On the other hand these subdivisions and other fields of scientific disciplines has to correlate with each other and thus interdisciplinary researches were necessary. For example, since biological phenomena are usually non-linear and discontinuous, biology has been the science most remote from mathematics. Rather recently, however, topology has been applied on some fields of biology and interdisciplinary science such as theoretical ecology developed.

The tendency of the interdisciplinary research reflected on science education inevitably. Interdisciplinary courses had been provided in many Japanese colleges, and biology became one of the important fields to support those interdisciplinary courses. In 1973 the writer sent questionnaires to 75 national universities in order to research on practical status of the interdisciplinary subjects. It was most surprising that more than two thirds of the universities had already provided the course even only two years after the amendment. It has also been clarified that around two thirds of the courses are related to scientific fields. Moreover, biology occupies the main part of those scientific fields. Accordingly, it is not too much to say that biology has a most important role in liberal education. One of the important aims of liberal education is character molding or character training in order to adapt people to their future community. From this point of view, biological education is destined to be closely related to development of the community.

There is a very important condition, however, to make about interdisciplinary science. Fundamental science must be acquired distinctly. In the same way, biological concepts and knowledges should be acquired by learners before they study an interdisciplinary course within which biological sciences are incorporated.

II. Sociobiological Education

When we try to make biological education accord with community or society, sociobiology evolves as an interdisciplinary science. The following table shows an example of the sociobiology which was organized by the present author and hundreds of students enrolled in this course during 1972-74.

Table 1. Outline of "Life of the Human Being" as an Interdisciplinary Science

Chapter 1 Past and future of human beings

1. Living materials and life.
2. Evolution, from prokaryotic and eukaryotic to man.
3. Characteristics of man.
4. Prosperity of man and ecosystems.
5. Human activities and morals.
6. Cooperative activities between man and machinery system.
7. Future of information society and roles of man.

Chapter 2 Life
1. Life in medical sciences.
2. Life in the law.
3. Lifo in philosophy.

A deficiency of an interdisciplinary course is that several teaching staffs have to be assigned to the class usually, since it is very difficult to find an all-around lecturer covering so many disciplines. On the other hand, if several lecturers might be assigned for one class one after the other, the tendency is that they just peddle out a lot of knowledges in pieces. In order to compensate for this defect, the number of lecturers should be as small as possible, and also a core program and its text-books have to be prepared.

Environmental education may be one of the most popular and effective subject of the interdisciplinary sociobiological course, because it can easily be taught by one biologist, especially an ecologist, and also many textbooks and references are available in any country. Principles of environmental science authored by E.F.Watt (Mc-Graw-Hill Book Company) is an example available in world wide.

Another effective course is based on a minicourse or module system. The system is most easily organized because the topics are not linked systematically, and specialists of each scientific topic can attain most effective educational results very easily.

III. Human Biology

Since "Man" is a group of organisms and their living phenomena are similar to those of animals, human biology is more attractive to study for many students and pupils than to learn biology as a pure science. Of course Man is a peculiar animal and has a special status in ecosystem, the peculiar characteristics are most worthwile to study in order to recognize the biological uniqueness of humans. The following table shows examples of contents of human biology texts.

Table 2. Contents of Human Biology authored by Dr. T. Kikuchi

1. Biogenesis. Origin of life; living materials; chemical evolution; metals, salts and water.
2. Climates and adaptation of Man. Why human temperature is around 37°C; human races and circumstances, biorhythms.
3. Energetics. Body size and energy exchanges; mechanism of heart and blood.
4. Genetic information. DNA and amino acids; episome and virus; regulation by gene action, genetics of behaviour.
5. Immunology and allergy. Protection from invasion by microbes; Ig; allergy; blood types; transplantation of internal organs.
6. Ages of the bubonic plague. Leprosy; pest; syphilis and Columbus.
7. Historical progress of vaccines. Pox vaccine discovered by Jenner; Pasteur and rabies vaccine; yellow fever vaccine; polyomyelitis.

8. Salvarsan to penicillin. Syphilis and salvarsan; discovery of sulpha drugs; drugs from moulds.
9. Cultural history of malaria. Infected air of "Mal-aire"; intermediator of malaria; quinine from Peru; sickel-cell anemia; ariddle of Hindu.
10. Hormones. Insulin; sex hormones; using the pill.
11. Problems on cancer. Introducing factor of cancer; abnormal metabolism by cancer; cancer and virus; molecular biology of tumour virus.
12. Poisonous chemicals. Toxic symptom of Hg, Pb and Cd; thalidomide; insecticides and pollution; poison gas.
13. Mental disorder. Psychiatry; stress; mental hygiene.
14. Hereditary disease. Chromosome aberration; gene aberration; in breeding and hereditary disease; civilized society and hereditary disease.
15. Population and food. Prediction by Malthus; Vitaminosis; limited amount of foods; population capacity of the earth.
16. Origin of cultivated plants. Birth place of cultivation; characteristics of cultivated plants.
17. Fortune of domestic animals. Birth place of domestic animals; characteristics of domestic animals; future of domestic animals.
18. Fermentation and foods. Wine and alcohol; baking bread; native land of cheese; soy-beans culture, protein from petroleum.

Subject of biology of the upper secondary level is strongly controlled by the Ministry of Education, and it is very difficult to reform it to human biology in the present situation. But slightly reformed curricula such as "biology oriented to human beings" are plentiful. The famous textbook "Biology Today" by C R M Books, is an example of this approach, though this is college level. In recent Japanese textbooks of the secondary level many biological facts and phenomena are explained dealing with humans. Especially human beings are emphasized on fields of developmental biology, regulation systems, genetics and ecology. Biological facts and scientific aspects of biology are becoming minimized whereas biological content concerning human beings and their society are getting to be significantly considered.

KUWAIT

BIOLOGICAL EDUCATION AND COMMUNITY DEVELOPMENT IN KUWAIT

by M.S. Subbarini

Introduction

The State of Kuwait lies between latitudes 28° 40' and 30° 10' North and between longitudes 46° 30' and 48° 30' East. It is bound on the North and North West by Iraq, on the South-West by Saudi Arabia and on the East by the Gulf. The State of Kuwait consists of a 17818 square kilometers triangular area of desert terrain sloping gently towards the Bay of Kuwait. Of this area 1000 sq.kms. represent ten offshore islands.

Following the oil discovery in 1946, Kuwait grew very fast. The urbanization of the country has been very rapid and Kuwait has changed to a modern country in just under two decades. The population of Kuwait has also increased rapidly. While the total population in 1944 was 70,000, it increased in 1975 to 991,392 of which the total number of Kuwaiti nationals is 470,121 and the number of non-Kuwaitis is 521,269.

We, in Kuwait, believe that biological knowledge should be taught for the benefit it brings to the quality of life and to human welfare in the community, and the effect of this knowledge in personal and in social decision-making (skill of decision making). No survey or study was made to ascertain the degree of success attained in this respect, since our approach to "New Biology" is relatively new.

So, my intention is to point out the major formal and informal locations which do play a part in community based education.

A future look to the role of biology education in community development will be mentioned on the basis of plans and expectations for Kuwait in the coming decade.

Scope

In presenting the status of Biology Education and Community Development in Kuwait, it is understood that the role of both formal and informal education should be covered, since they are both influential in supporting community development.

The formal education locations include all the educational institutions, for both pre-university education and tertiary education. While among the informal education locations, that deal in some respect with biological knowledge, are: the Science and Natural History Museum, the Zoological Garden, the Professional Societies, the Summer Clubs, the Science Club, petrochemical and other industries ... etc.

Although the community agencies at both formal and informal levels, seem to be numerous, yet schools are the most influential in community development. This is due to the fact that around fifty percent of the total population of the country are enrolled in schools, and also that the rest of the agencies and localities are more or less newly established.

So, my stress and emphasis will be given - in this paper - to the role of the pre-university Biology Education on community development.

Formal Education Locations in Kuwait

The formal education locations include kindergartens, the schools of general education, the adult education centres and the tertiary education institutions. Each educational institute has a kind of influence in supporting community development through biological curricula and the activities connected with them, as seen from the following survey:

Kindergartens. Children (age 4-5) are admitted to official or private kindergartens. Kindergartens run curricula supervised by the Ministry of Education.

Kindergarten syllabuses use biology knowledge, about local plants and animals, as a tool for developing the desired skills and experiences at this level. Children, also, practice activities that effect health attitudes such as good habits, proper use and care of teeth, cleanliness etc. This role of kindergartens is so important, since a considerable number of children cannot learn such attitudes (or learn in a wrong way) at home due to ignorance or carelessness of parents. Furthermore biological phenomena in the environment are the first and most appealing things that arouse the curiosity of children.

General Education Schools. The general education system is composed of three stages, four years each, as follows:

Stage	Number of years	Age of pupils
Elementary	4	6 - 10
Intermediate	4	10 - 14
Secondary	4	14 - 18

The number of hours per week is less in elementary schools than the intermediate and secondary schools. Also girl schools have more hours due to an extra course in Home Economics.

Stage	No. of hours per week	
	boys	girls
Elementary	34	35
Intermediate	37	39
Secondary	37	39

Biology in the pre-secondary (primary and intermediate) is integrated with other science disciplines in a form of integrated science courses. In the primary stage the core of the integrated science course is man and the universe. While, in the intermediate stage, the core is man and matter (1st and 2nd intermediate classes), and man and energy (3rd and 4th intermediate classes).

Elementary School Biology as an integral part of the science course requires that pupils should be able to achieve the following goals:

(1) Understand how and why houses should be kept clean and hygienic.

(2) Observe and study the main groups of living organisms existing in the environment with a view to being able to recognize the main animal and plant groups, percieve their main features and distinguishing characteristics. This skill should result in the pupil becoming more aware of nature, and with increased awareness, there will be increased joy in the beauty of nature and subsequently more concern for it.

(3) Understand in a very general way the main problems facing a living animal and a living green plant, and the interrelations, interaction and interdependence of all living organisms in the environment.

(4) Understand that human beings are "animals", with preliminary understanding of bodily structures and functions.

(5) Understand the role of food in the human body and to be able to relate that to food problems, especially malnutrition in the broadest meaning and the phenomenon of hunger causing so many deaths in many parts of the world.

(6) Understand in a preliminary way the concept of reproduction, with special reference to the population explosion problem.

(7) Understand in a general and preliminary manner the effect of "familiar" pollutants on the environment.

(8) Reject popular myths related to the behaviour of some animals, such as being pessimistic when hearing the sound of an owl or a crow or seeing a black cat at night ... etc.

The biological knowledge in the primary schools is obtained mainly through observation, investigation and a variety of curricular and extra curricular activities.

We feel that Elementary Biology helps the pupils to begin the development of responsible attitudes towards themselves, other people, familiar animals and all other creatures in the world of life surrounding them.

But how does Elementary Biology affect the community? The pupils themselves are affected, and they are part of the total community. It is also expected that the biological knowledge gained by pupils will affect parents and adult relatives who may not have had any experience of such knowledge. The role of the "pupil" becoming a teacher to adult relatives has been evidenced and established. This was the reason for a T.V. program called "Science and Life" aimed at the layman which was launched 15 years ago. It helped in giving credibility to what pupils conveyed to parents.

Intermediate School Biology aims at enabling pupils to achieve the following goals:

(1) Understand that earth is a planet suitable for sustaining and

maintaining life.

(2) Understand in a general way the relationship between other organisms and health: the kinds of organisms that provide food to man; the importance of being clean and careful to avoid or combat organisms that can be harmful.

(3) Understand the integration of body systems' functions. This will help the pupil to be aware of the intricacy of being healthy.

(4) Understand and recognize the importance of the natural balance in ecosystems. This might help in developing awareness towards the environment.

It is clear, now, that the Intermediate School Biology stresses and emphasizes Health Education (including sex education) and Environmental Education. As for sex education, it was a taboo in the fifties and early sixties of this century. Now, it is dealt with in the 3rd intermediate class (age 13), but in a general and cautious manner.

As to the influence of Intermediate School Biology on community, we believe that this influence is enhanced by the fact that pupils are somewhat older and hence capable of arguing with elders more convincingly.

In Secondary Schools, Biology is studied separately from other science courses as follows:

			hours / week			
Subject	1st Sec.	2nd Sec.	Science Divi.	Arts Divi.	Science Divi.	Arts Divi.
General Science	-	-	-	2	-	-
Biology	1	2	3	-	4	-
Chemistry	1	2	4	-	4	-
Geology	-	1	1	-	1	2
Physics	2	2	5	-	6	-

Secondary School Biology adopts three approaches: 1) the descriptive approach, 2) the scientific enquiry approach and 3) the community approach. Most emphasis is laid on the first two approaches, but in future projects we intend to shift the emphasis more towards the community approach.

At present Secondary School Biology aspires to students achieving the following goals:

(1) Recognize the diversity in living things.

(2) Comprehend that man is built up and functions according to the same principles as other living creatures - unity of living things.

(3) Account for those factors in the environment which are of importance to man's health, and on the basis of his general knowledge relative to this, analyse consequences of new influences on health (priority is given to the influences which emerge in the students daily surroundings).

(4) Show a responsible attitude towards problems concerning man's interaction with environment.

(5) Explain the most important genetic laws, and use these in examples concerning genetic inheritance in man.
(6) Judge the dependence of human behaviour patterns on genetic and environmental factors.
(7) Analyse problems concerning human ecology, its position in relation to nature and therefore man himself, bearing in mind man's position of power and his capabilities for intervention and the consequences of these actions.
(8) Recognize biological problems and formulate hypothesis concerning their solution.
(9) Suggest and carry out experiments and/or observations to illustrate the formulated hypothesis.

In view of the above mentioned goals, human biology is emphasized. Also topics which have direct bearing on the community are emphasized, such as: health, environment, nutrition, population, genetics and behaviour.

Talking about Secondary School Biology is incomplete without mentioning the "credit hours system" which is on trial now in one boys' and one girls' secondary schools. This system allows a variety of options and approaches for students, also it encourages them to use discovery and problem-solving approaches and the project method for self-teaching.

In the credit hours system the four years of the secondary school are divided into eight semesters, 15 weeks each.

In the first two semesters, students study common courses including two integrated science courses - Man and Environment, and Man and Energy -. After that students choose one of the eight streams offered by the school. In the science stream, students have to study one biology course - Metabolism and Homeostasis in living things - , and have access to another course - Continuity of Life -. There are, also, two biology courses accessible to the science stream students as well as students in other streams. These two courses are: "Biology of Human Populations" and "Science for Life".

The Biology courses in the credit hours system emphasize all that has been mentioned before about biology in the regular secondary school. In addition to this the following goals are expected to be achieved by students:

(1) Account for the morphological, physiological, and psychological development of a person from a fertilized egg to old age.
(2) Explain the use of the most important means of contraception, their advantages and disadvantages, and their effects.
(3) Account for man's phylogenetic development and those mechanisms which are of importance for this development.
(4) Describe how man through his phylogenetic development has influenced the environment, and on the basis of this knowledge, assess how man's influence of the environment will develop under different conditions.
(5) Judge critically explanations of the biological traits in man,

adaptations to environment and way of life in pre-historic time, on the basis of available knowledge about pre-history and the general theory of evolution.

(6) Assess the biological basis for the social problems of our times, their consequences and explain the assessment on the basis of biological research results.

It is evident that biology education in the credit hours system is leaning heavily towards the community approach. It is the first time in Kuwait, that students have the chance to study about contraceptives, evolution of man, human races, and genetic counselling.

In brief, biology in this system isemphasizing the environmental, nutritional and sex educational concepts with indications of what is relevant to present and future human populations.

Also, it is the first time that students have the chance to study local case studies. "Food in Kuwait" is one of the main case studies which are introduced in this system. The importance of this case study appears when we know that Kuwait imports 80% of its food, and a variety of unhealthy food habits are still rooted in the population. These naturally lead to health hazards. Obesity for instance is an embarassing problem in the country. Similarly a-vitaminosis.

Biology Education at the Tertiary Level. Tertiary education in Kuwait can be divided into three categories: the university, the centre for community development and continuous education and the vocational institutes.

Kuwait University, like most of the universities in the developing countries, teach the same biology topics of the pre-university level in more detail. Also, research of academic nature has been inducted at the university. Some of the research results took their way to the university and pre-university curricula, which are the major media for development of community education.

The Centre for Community Development and Continuous Education is a specialized body in Kuwait University. It is actually a location of community focussed education, where people from all walks of life can be enrolled in special courses. Until now, very few biology related programmes have been introduced, but the future is promising.

The vocational institutes are specialized locations where students are trained for a special vocation. The Teachers' Training Institute qualifies students to be teachers in primary (Elementary) schools. The Health Institute for girls qualifies girls to be health advisors in secondary schools. The Nursing Institute qualifies girls to be nurses. The Agricultural Institute (which has been terminated now) qualified students to work in extention programmes of forestry, pastures and agricultural developments.

All the above mentioned vocational institutes admit students on passing the Secondary Education Certificate Examination. Students spend two years in such institutes and graduate with a diploma.

The graduates of the vocational institutes, of course, communicate the Biology knowledge pertaining to their vocation to the sector of people with whom they come in contact.

Informal Education Locations in Kuwait

Although the formal education locations in Kuwait play the biggest role in community education, yet there are some informal locations and activities that relate Biology knowledge to community development.

The locations and activities that deserve mention in this respect are:

(1) The Science and Natural History Museum: Besides preserved animals, this museum contains a special hall for health education accessible to the public. Visitors can watch films and operating models about the physiology of the functioning human body. Models representing common diseases are also exhibited. Seven hundred people visit the museum daily. This is besides the programmed visits of school students.

(2) The Zoological Garden: The Kuwait Zoo is one of the main recreation spots in the country. It includes a variety of animals from all parts of the world. Hundreds of people visit the Zoo daily. There, they have a chance to watch animals previously only seen in pictures and get acquainted with them.

(3) The Professional Societies: More or less, each profession has its own society, Teachers' Society, Engineers' Society, Medical Society, Agricultural Engineers' Society, Geographical Society, Environmental Protection Society, etc ...

All the professional societies are patronized by the Ministry of Social Welfare. Each society takes part in the community development in its field of speciality. Societies usually publish special papers, invite for seminars and give advice and suggestions to the government.

The Societies that relate Biology knowledge to community development are: the Medical Society, the Agricultural Engineers' Society, the Teachers' Society, and the Environmental Protection Society.

The Medical Society co-operates with the Ministry of Health and Kuwait T.V. in health education programmes. Also this society runs a variety of activities on the annually observed Health week (posters, seminars, special programmes in Radio and T.V. ... etc.).

The Agricultural Engineers' Society sponsors a prominent activity each year. It organizes an "Agricultural Exhibition". Visitors have the chance to learn how to cultivate plants and can have seedlings free of charge. Plants in Kuwait are valuable and need special care if they are to grow and remain alive. The Agricultural Engineers' Society gives free of charge advice to any body who requests such advice.

The Teachers' Society participates in the High Committee for the Development of Curricula in the Ministry of Education. This committee is the highest authority in the country for curriculum development. The Teachers' Society also organizes an "Educational Week" each year in which it evaluates a topic related to community development. Biology education is one of the major topics that receive attention in this respect.

The Environmental Protection Society (EPS) plays a prominent role in promoting public awareness about the environment. It organizes each year an International Scientific Exhibition for the protection of the envrionment. Also it participates in the organization of an Environmental Week each year.

The above mentioned informal locations still have much to do for community development, other locations may have a role in relating Biology knowledge to community development, of these are: the Petrochemical and other Industries, Summer Clubs of the Ministry of Education, the Science Club, the Kuwait Institute for Scientific Research (KISR) ... etc.

(4) The Television Medium. It has been recognized all over the world that television has a strong impact on the community. This effect and impact could be utilized most effectively in furthering, enhancing and broadening the influence of formal institutions. Already there are many very good T.V. programs produced internationally which deal with biological knowledge in general and environmental problems.

Kuwait T.V. televising on two channels makes use of very many of these programs subtitling them in Arabic.

However, a special telecast is also produced under the title of "Science and Life". This 30 minutes program which started in 1963 is still running strong and is telecast once weekly. It has been so popular that it is televised, in addition to Kuwait, in Jordan, Dubai, Abu Dhabi, Saudi Arabia, Morocco, Tunisia, Sudan und Qatar.

The program encompasses all subjects to science and life, environment, health, interaction of living and nonliving things in various ecosystems all over the world, physiology and physiological problems, recent scientific advances and discoveries, genetics and genetic engineering etc....

Future Biology Education

It is well recognized that Biology as a science has a more direct and more profound meaning for man than do the physical sciences. Also it is well recognized that biology is now dominating the scene of scientific achievement taking this role over from physics. The advances in genetics, neurobiology, immunology ... etc. have opened the way to the solution of many human problems, and as usual in science, created many more.

Since, it is obvious that the progress of Biology is of great social importance, so Biology Education should play the role of the interpreter of biological concepts, theories and researches, and serves as the intermediary between scientists and the lay public and especially the decision-makers. The latter is of special importance since serious problems may be related to the widespread ignorance of biological concepts among leaders and decision-makers in both the developed and developing countries.

So, Biology Education as a basic component of a liberal education system in the future should:

(1) Be taught in a social perspective and within a human context, perhaps as a science of Homo sapiens.
(2) Include values and ethics as goals, recognizing that there are a moral and aesthetic as well as scientific answers to human problems.
(3) Emphasize social issues man will have to face, such as: the preservation of human health, the respect of human life, the balance between individual rights and the needs of society, the nature of human differences etc....
(4) Have courses organized more according to biosocial events and problems that have meaning for the quality of life, rather than according to the logic of biological disciplines.
(5) Encourage additional cognitive skills, such as decision-making, evaluation processes, knowledge validation, problem resolution, concept of risk and ecological thinking.
(6) Orient Biology Education towards the future, giving the different sectors of the community opportunities to consider various alternatives for the future course of human efforts.

In the light of the above mentioned futuristic outlook of Biology Education, educators in Kuwait are taking steps towards the development of strategies for a total community activity where the school will still be the major community agency concerned with community development. But the efforts of the school will have an overall impact beyond its borders. This role of the school necessitates that the biology curriculum should not be static (although the dictionary definition of curriculum means a fixed series of studies), since we live in an age of rapid change, and a static curriculum soon falls behind the times.

Continuing education also has its location in the total community activity, in addition to the informal agencies whose activities are expected to play a considerable role in the national movement towards community-based education.

In brief, the future of Biology Education in Kuwait will not be a mere teaching of "Biology Knowledge" but it will be formulated in a way that will help people to be aware of problems facing them and to get accurate and relevant information, as well as creating a sense of responsibility that would drive people to participate in community affairs.

Conclusion

I hope I succeeded in presenting the aspects of Biology Education in Kuwait. It can be stated that the community approach in Biology Education is in a stage of development and that there is plenty of room for improvement.

In future, biology curricula will be oriented towards the community needs, continuous education locations will be extended and increased in number, the informal education agencies will have a definitely determined role in community education, and the lag between science research and Biology Education will be overcome.

It is hoped that, through cooperation of all biology education locations, an effective development of biology education will be the out-come. This in effect would mean a better quality of life to the community at large.

Bibliography

BSCS Journal, Volume 1, No. 4, Nov. 1978.

BSCS Journal, Volume 2, No. 1, Feb. 1979.

Final Reports of the Main Committee for the development of Science Curricula - Ministry of Education - Kuwait - 1976, 1977, 1978, 1979.

The General Objectives of Education in Kuwait as formulated by Ministry of Education in March 1975.

A Summary Statement on National Science and Technology Policy in Kuwait - Planning Board, State of Kuwait, Sept. 1975.

SUBBARINI, M.S., The status of Biology Education at the General Education Level in Kuwait - A report submitted to the Regional Meeting to develop a Teachers Source Book on the Biology of Human Populations and their environment, organized by UNESCO at Yarmouk University, Irbid - Jordan, Jan. 28 - Feb. 2, 1978.

SUBBARINI, M.S., the status of Science Education in Kuwait. A report submitted to the Regional Seminar to develop Science Education in Arab Gulf States, organized by Qatar University, Doha, State of Qatar, March 10 - March 19, 1979.

MAURITIUS

THE BIOLOGICAL SCIENCES IN MAURITIUS

by M. Atchia

Introduction

This paper examines the position of Biology in the school curriculum and the way in which this position has changed during the past 15 years.

During this period three main trends have been noted:

(a) An expansion in the number of students taking Biology.

(b) A diversification from the classical Biology/Zoology/Botany situation to one where several bio-social, integrated or applied subjects have been introduced.

(c) A very clear evolution towards the utilization of the active method of science teaching, with a growing accent on ecology and environment.

At the Primary Level

The six years of schooling (from Std I at 5+ years of age to Std VI at 11) are uniform within the 250 primary schools in the island. Compulsory subjects in the curriculum have traditionally been (are still in 1979!) the following:

- English
- French
- Geography
- Arithmetic, now Mathematics

Optional subjects include

- Singing
- Games
- Drawing
- religious education and a choice of 1 out of the following languages: Hindi, Urdu, Arabic, Marathi, Telegu, Tamil, Mandarin.

Several recent developments have tended to modernise the curriculum but in no way lessen the burden, specially in the case of children who opt for 1 oriental language and have therefore three languages to study. New trends have includes:

(a) the gradual replacement of Arithmetic by modern Mathematics - this move is certain to help in the introduction of Sciences because of the closer links between the processes of modern Maths and Science;

(b) the creation of school gardens and of gardening, an excellent idea which has two objectives:

(i) familiarise children with common local plants,

(ii) give them a taste for and develop skills in food production.

However, less than 50% of the schools have a decent and functioning school garden, either from lack of space, or lack of effort from school authorities. Moreover, when successful, a garden can become "the most impressive undertaking in the school". [1)]

(c) The introduction of Primary Science. This is still in the form of a pilot project in about 100 schools and since its inception in 1974 has gone under several names:

Primary Science — Health & Environment
Junior Science — Etude du Milieu

and finally, today, Environmental Science.

The present syllabus of Environmental Science which has been circulated to all schools, is largely biologically based. It is a spiral syllabus covering three years work (Std IV - VI); the main components,

1) Science Radio Broadcasts for Primary Schools - an investigation into the impact of these broadcasts on pupil's attainment. M. Atchia, MIE Occasional Paper (1977).

repeated each year, are:

1 AIR
2 WATER
3 COMMON MATERIALS
4 PLANTS IN YOUR ENVIRONMENT
5 ANIMALS IN YOUR ENVIRONMENT
6 OUR BODY AND HEALTH

A series of TV broadcasts for primary school teachers and accompanying radio broadcasts for pupils have been operational for several years; these are prepared by a team of lecturers from the Mauritius Institute of Education and the primary teachers training college, primary school inspectors and several primary school teachers under the co-ordination of this author. An evaluation of these radio broadcasts for pupils conducted by Atchia [1] has shown that:

(i) Radio broadcasts of the type utilised have a significant role in helping Std IV pupils of Mauritian schools in learning science.
(ii) An increase of about 20% in attainment was noted as a result of utilisation of radio (20,2 % in rural areas, 18,9 % in urban areas).
(iii) Radio broadcast may be generally useful in helping young children in learning science.
(iv) A small but significant difference exists in pupils attainment in science at Std IV level between purely rural areas and urban areas.
(v) Pupils attainment in urban areas were about 7,8 % better off than rural areas.
(vi) The absence of a radio in schools is of slightly less consequence in urban areas than in rural areas.

A pictorial objective test was devised by the author as instrument of evaluation to measure the achievement of students in selected topics of primary science at Std IV level.

Finally, it is to be noted that by 1982 when Environmental Science will have been established in Std IV, V and VI of all primary schools, a merger with Geography and Social Studies is envisaged to form a new integrated subject, Environmental Studies. This subject will then replace Geography in the primary school leaving certificate examinations (CPE) still a very important feature of primary school life as it is utilized as an instrument of selection for entry into secondary schools.

At the Secondary Level

During the first two years of secondary school (age 12 and 13) students take 5 core (compulsory) subjects:

- English
- French
- Mathematics
- Social Studies (integrated Geography, History, Economics).
- Integrated Science (partially integrated Chemistry, Physics, Biology).

[1] Science Radio Broadcasts for Primary Schools - an investigation into the impact of these broadcasts on pupil's attainment. M. Atchia, MIE Occasional Paper (1977).

together with several optional ones:

- Home Economics
- Visual Arts
- Industrial Arts
- Commercial Studies
- Agricultural Science
- One oriental language.

The Integrated Science course offers a balanced 'diet' of the three basic sciences, with an accent on measurement and experimentation right from the start and an accent on environmental matters later on. The main aims of the course are: imparting knowledge, achieving comprehension, developing skills and creating a positive attitude in the students. This is achieved through the use of the discovery (active) method throughout and on the study of 16 main themes which are:

THEME 1 OBSERVATIONS AND EXPERIMENTS
" 2 LIVING THINGS IN YOUR ENVIRONMENT
" 3 THE AIR AROUND US
" 4 WATER WATER EVERYWHERE
" 5 SUPPORT AND MOVEMENT
" 6 YOU AND YOUR ENVIRONMENT
" 7 NATURAL RESOURCES
" 8 POLLUTION
" 9 CONSERVATION PROJECTS
" 10 POPULATION
" 11 ENERGY AND ENERGY SOURCES
" 12 INTRODUCTION TO ELECTRICITY
" 13 APPLIED ELECTRICITY
" 14 ENERGY AND SUBSTANCES
" 15 BUOYANCY
" 16 PRESSURE.

The Integrated Science Course of Forms I - II of our secondary schools has gained wide acceptance and is now studied in all 156 secondary schools in the island. A conclusive piece of research, connected with this course, was conducted by Atchia [1] in 1977-1978 and showed that "it is possible to modify attitude towards the environment and bring it nearer to a fundamental ecological outlook", through the experimental study of environmental themes. A very large change in environmental attitude was recorded; the % change (towards a positive environmental attitude) was much greater in pupils from "B" [2] schools (96,4 % significant difference) than in pupils from "A" schools (58, 3 % significant change).

During the third year of secondary school (Form III) students take three courses:

---

[1] Concepts and dynamics of environmental education, with particular reference to Britain, Africa and Mauritius. Ph.D. Thesis, University of Salford, 1978.

[2] (In the Mauritian context quality of school = status (A or B) = pupils ability + teachers qualifications + support facilities).

Matter and Life — Matter and Energy
Matter and its Reactions,

i.e. basic Biology, Physics and Chemistry. The material has been written, trialled and published by the Mauritius Institute of Education (as was the Form I and II material); while retaining the discovery approach it provides some more basic information and structural patterns (e.g. classification of living things, main physiological processes, types of substances and of chemical reactions, major physical laws etc).

Thus Chemistry, Physics and Biology are now on an equal footing up to Form III so that students in Forms IV and V will in future have equal incentive in choosing 1, 2 or 3 of the science subjects offered at School Certificate level; the subjects which a biologically orientated student may take are:

Biology — Human and Social Biology
Botany — Agricultural Science.

Many students still prefer to take the 3 basic sciences(i.e. Chemistry, Physics, Biology) leaving specialization for later. This situation contrasts markedly from the one obtained prior to 1966 when Biology was the very poor relative in the family of the sciences.

The major factors helped to change the future:

(a) the UNESCO Biology Project (66-71) for the improvement of the teaching of Biology in Africa (several young Biology teachers on the island got involved in the project supported by personnel and literature sent from UNESCO);

(b) the coming to Mauritius for a few months of 1968 of Dr. Rex Meyer from Macquarie University, within the context of theabove named project. The seminars run by Dr. Meyer contained novel ideas of curriculum development, and of new approaches to Biology teaching (the Nuffield and B.S.C.S. approach); these helped to boost Biology education in Mauritius in such a way that it has never looked back since.

The number of students taking Biology at School Certificate level is given below. This examination is taken at 16; candidates are entered for 6, 7 or 8 subjects, English Language being compulsory.

Year	Total No. of entrants for School Certificate in Mauritius	Entries for Biology No.	: % of total
1955	1605	90	5,6
1964	3634	240	6,0
1966	5251	390	7,5
1973	9071	1408	15,5
1974	9262	1646	17,8
1975	9899	2289	23,1
1976	10862	2501	23,0
1977	11664	2789	23,9
1978	12621	3000	23,8

A look at the number of students taking science subjects at School Certificate in East Africa and Mauritius for the period 1960-64 showed that in the former area Biology was the overwhelmingly predominant science subject, with few candidates taking Chemistry and fewer still Physics, whereas in Mauritius Chemistry was more widespread, Physics next and Biology the least studied of all sciences. Being given that the 3 East African States and Mauritius are all agricultural economics it was anomalous that biology was so unpopular here.

The Future

In 1968 Rex Meyer suggested to the 21 biology teachers on the island (we are now over 200) that we should take care to avoid a purely academic programme and that the biological sciences "should be demonstrably human in orientation, stressing the biology of man and his relationships with the environment", as well as "featuring prominently the special health, hygiene, population, agricultural and conservation problems of Mauritius". [1]

As far as the biological component of Forms I - III programmes of study is concerned, this injunction has been fully achieved; it is being slowly attained at the upper secondary level. Thus the biological sciences of today will serve the future through the thoroughly relevant training they offer in a country which for a very long time more will remain essentially agricultural in vocation.

MEXICO

BIOLOGICAL EDUCATION AND COMMUNITY DEVELOPMENT IN MEXICO

by J.M. Gutierrez-Vasquez

Introduction

Diagnostic research studies on the quantitative aspects of the Mexican educational system have yielded enough information to formulate specific programs for action. Unfortunately, this is not the case for the qualitative aspects, the study of which has been initiated systematically just recently in a few centers for educational research. In spite of this deficiency, and taking as a base the quantitative information, it is possible to draw some preliminary conclusions about the quality of education in our country. For example: the indexes of failure to pass or to be promoted to the following grade are very high all along the system (in elementary school 17,0 % fail in 1st grade, 11,6 % in 2nd grade, 9,9 % in 3rd grade with an average of 10,7 % failures for the grades 1 through 6. The drop-out rate is also very high (in the elementary

[1] Ministry of Education and Cultural Affairs, Mauritius. UNESCO biology project for Africa. Memorandum to the Executive Committee of biology teachers, 1968.

school 6,2 % drop-out in the 1st grade during the academic year, 4,9 % in the 2nd grade, 5,1 % in 3rd grade, with a drop-out rate of 5,4 % for the elementary school system as a whole in a single academic year). For every 100 children registered in 1st grade elementary school, 75 are no longer on the graduation lists at the end of junior high school; in fact 54 of them drop out before ending the elementary school, which is compulsory by law.

These rates are similar in the majority of the under-developed countries. Nevertheless, their magnitude is such, that it is unwise to advocate single causes to explain this situation or to simply blame factors external to the educational system, as the social and economic conditions of the children who fail or drop out. Preliminary results indicate relevant internal factors such as: meaninglessness of contents to everyday life and needs; inadequate methodology of the learning/teaching process; erroneous criteria, procedures and instruments to evaluate academic achievement of pupils; inadequate preparation of teachers; excessive non-teaching tasks imposed on teachers; low salaries, compelling teachers to look for a second job, inside or outside the system; virtually non existent in-service teacher's training programs; and the administrative and power structure of the national educational system.

Contents of Education and Community Needs: a Disfunctionality

There is a gross disfunctionality between the contents of education and the needs and problems that everyday life poses to pupils while they are in school and after they leave the system. The educational contents form a part of the school culture that, in terms of the explicit or official curriculum, has little or nothing to do with the demands or the culture of the community of which the school forms a part. The culture of the school is useful only if one remains in the school system, and in this sense is markedly propedeutical in nature: 1st grade elementary school is useful just for those going on to 2nd grade; elementary school itself educates just for going on to secondary education; and this is good only for going on to the university system. Only in the light of hidden curriculum considerations does the educational system show its logic and its functionality with respect to the demands that society poses to to the school as a socializing institution.

A Shift towards the Study of Relevant Problems: Elementary and Junior High Curriculum Reform

As part of a vast program of educational reform, in 1971 the Ministry of Education set up several projects to produce new curricula for all of the courses of study both at the elementary and the junior high school levels. New curricula and new educational materials were produced from 1972 to 1976. By request of the federal government, our group (Department of Educational Research (DIE) of the Center for Research and Advanced Studies of the National Polytechnic Institute) was responsible for the science education area.

The aim of the new science curricula is focused not only on the understanding and intelligent use of the processes of nature, but also on the understanding and intelligent use of ourselves as human beings. Science is conceived of not only as knowledge, but also as methods and procedures with which to obtain and build new knowledge; so, science is presented as an open, endless activity. Through the curriculum, science is presented as part of our culture with both practical and conceptual implications. There is the intent to generate attitudes against consumerism and tasteless advertising; and also, by stressing the principle that natural phenomena have natural explanations, to liberate children from the fear, tension and anguish generated by poor education and the communication media. The curriculum is also aimed at preserving and developing some characteristics that children have in common with scientists: to question, to criticise, to participate, to do, to create.

The concept of learning which is utilized implies that to learn is to modify our way of being, our way of doing and our way of thinking according to the knowledge acquired; and so, the curriculum is based on methods, procedures, ways and means. Knowledge is obtained by the interaction of observation, experimentation and human intelligence. "Truth" is only a prevailing certainty and, as such, must always be submitted to verification, must always be proved.

The up-dated contents are structured around main concepts, general ideas, fundamental principles and basic or persistent problems. The children get into the content through their own activity and participation, working and investigating the world of reality both in teams and individually in order to develop particular abilities, aptitudes, skills and attitudes. There are six guides for the teachers as well as six books for the children.

Examples of problems relevant to community development included in the new curricula for elementary school are: the vegetable garden, the animal farm, plant nutrition, animal nutrition, the corn, erosion, the composition and structure of the soil, microbes, health and disease, agriculture, husbandry, the water cycle, ecosystems, fisheries, forestry, mining, pollution, human nutrition, genetics, population, environmental education and sex education; for junior high school: ecosystems, environmental education, sex education, human nutrition, soil and erosion, minerals and mining, health and disease.

The science curriculum for the elementary school is national and implemented in various degrees all over the country. The junior high school science curriculum is experimental and is being tried out in a very small proportion of schools.

Biological Education and Community Needs: Three Examples.

In Mexico the research reports showing the correspondence (or lack of it) between the biological contents of syllabuses and curricula and the needs of the community are still to be produced. In this paper I do not pretend to present such a report because enough information has not been collected. Instead, I will present information on three projects, presently under way, in which biological education is structured taking into account the needs of the community as one of its main determinants. This kind of approach could be useful in helping to identify ways to link educational institutions with the communities to which they belong.

The first is the Instituto Nacional de Investigaciones sobre Recursos Bioticos (National Research Institute on Biotic Resources, INIREB), a research and development institution with educational programs ranging from the training of peasants to a Master of Science program in ecology and biotic resources.

The second is the Division de Ciencias Biologicas y Salud de la Universidad Autonoma Metropolitana-Xochimilco (Division of Biological Sciences and Health of the Metropolitan Autonomous University, Xochimilco Campus (UAM-X), in which the Bachelor of Science curriculum in biology is based in modules drawn from relevant needs ranging from the community to the national levels.

The third is a project in our Department (DIE): the Centros de Educacion Basica Intensiva para Ninos de 11-14 anos (Centers for Intensive Basic Education for Children 11-14 years old, or Project 11-14), in which science education and social studies (in fact mathematics and language also) are structured into a certain phase of the curriculum around a project of community health improvement.

In the first project the institution involved is a research center of the highest academic degree; in the second project the institution involved is working at the college level; the third project is an elementary school program designed for illiterate drop-outs.

The National Research Institute for Biotic Resources (INIREB)

This institution was created in 1974 in Xalapa, the capital city of the State of Veracruz (Mexico is a Federal Republic formed of 30 states), with several aims to fulfill through its programs, which include to study the biotic resources of the country; to combine scientific and technological research and development with educational programs in a diversity of levels in the same institution; to evaluate the impact of agricultural, industrial and economical development on the ecology of a locality or a region and their biotic resources; to decentralize research and development activities away from Mexico City, the country's capital city.

From its very beginning to the present time, the INIREB works out its programs mantaining an equilibrium between basic/applied research projects and development projects. The programs regularly take into account social, economical, cultural and political factors and implications.

Through their work, the INIREB regularly enters into contact with the needs and problems of peasants, fishermen, artisans, rural teachers and farmers working in small to medium communities, as well as those on engineers and architects working in larger ones. These contacts generate training courses given by INIREB staff members in the places suggested by the communities themselves. The courses offered by INIREB cover a wide range of interests and levels. Many of them are at the training level, as mentioned; others have higher technical goals; still others are given for university students and/or professors of college and post-graduate level, many of the courses having international participants. INIREB has also been a host institution for international meetings (International Union of Forestry Research Organization).

The knowledge of the needs of the communities on ecological problems and use of natural renewable resources, has generated not only educational but research and development programs also. Currently, INIREB Ecological Studies, Agroecosystems, Ecological Planning of Soil Use, Technological Development of Biotic Resources, Science and Technology of Wood, Utilization of Aquatic Biotic Resources, The Botanical Garden, and Support Services for Research. Each program encompasses several research and development projects, many of which are relevant to community needs (The Cinampa System, Regional Managing of Resources, Utilization of Wood, Development of Production Communities, Studies on Biotic Resources, Studies on Applied Agronomy, Management of Fresh-water Systems, Social and Economic Studies on the Management of Biotic Resources, and many others). These projects regularly generate formal and non-formal educational activities for the members of the communities involved.

INIREB publishes its results in the regular national and international specialized journals, and in its own publication services also: Biotica, a formal journal on ecology and natural renewable resources; Wood and Its Use in Building Industry, a technical journal; and many other periodical publications devoted to different sectors of the population.

The staff of INIREB is currently composed of 101 academic members, 94 technicians and 44 support personnel, for a total of 239.

Biology in the Metropolitan Autonomous University-Xochimilco Campus (UAM-X)

The Metropolitan Autonomous University was created in 1974 in order to alleviate the deficit in higher education for the ever growing capital city of the country. Of the three campuses in operation, UAM-X made itself conspicuous from its very beginnings by questioning the traditional concept of the university. Therefore, what is to be said about the B.Sc. in biology curriculum, holds true for the other curricula in UAM-X, including the rest of the sciences, the technologies and the humanities.

The biology curriculum in UAM-X is a modular structure in which each module is developed around a problem chosen through social practice. The students, working in teams of five, study the problem, define it, make use of scientific methodology to work out the problem in the field, and try to solve it or at least propose solutions for it. Through these activities, the students acquire knowledge and skills, and develop attitudes of social commitment.

The biological contents of the curriculum are not structured simply around knowledge itself (concepts, ideas, principles, theories, laws) but around relevant issues and problems. As a matter of fact, the biological contents are integrated with that of the other sciences (physics, chemistry, earth sciences) and with the social and economical aspects of the problem (module) under study. It must be pointed out also that each module tries to integrate research, teaching-learning and service to the communities. Each module implies the study and discussion of an extensive bibliography.

The curriculum offers four "areas of concentration": terrestrial ecosystems, hydrobiology, ecology and experimental biology. Every problem studied in a particular ecosystem (terrestrial or acquatic) is approached through the ecological or the experimental procedures and techniques.

At the end of the work for each module, the students deliver their results to the community in which they carried out their task.

Elementary Science Education and Community Needs: Project 11-14

As was pointed out at the beginning of this paper, the drop-out rate in elementary school is very high. This, in addition to the children that never go to school, is responsible for a figure of approximately 18 million people (out of a total population of 68 million) in the age range of 15-50 years that never completed their elementary education. By the end of 1978, the federal government requested our Department (DIE), to design and implement a new elementary school curriculum especially suited for urban drop-outs of 11-14 years of age all over the country (Project 11-14).

The new curriculum is being designed and experimented with at the same time. Ideas, procedures, contents, are discussed and then tested. Experimentation feeds-back the design team, so the curriculum is going to be the product of the work of the students, the teachers and the curriculum designers themselves; in addition, the members of the design team regularly teach in order to experiment themselves with parts of the curriculum. With this concept, Project 11-14 opened seven experimental Centers for Intensive Basic Education at the beginning of 1979.

These Centers are working in urban areas where elementary school drop-outs concentrate, generally squatter settlements characterized by very deficient urban services, high birth and infant mortality rates, high unemployment and underemployment, lack of amenities, etc. The Centers are meant to respond to the particular needs of boys and girls who are too old for the ordinary elementary schools and who have to work in order to somehow help augment the scant family budget. As can be seen, the time devoted to school and educational contents have become two of the main challenges to Project 11-14.

To begin with, the Centers offer the possibility of completing primary education in one third of the regular time (six years for ordinary elementary schools). Moreover, the Centers do not ask students to stay in school for four or five hours per day; instead, they run shifts of two and a half hours, and each student chooses the shift that best fits to her particular working or family situation.

The educational content of the curriculum is not viewed as an abstract or a résumé of the ordinary curriculum, but conceived of as the systematic practise of basic cultural skills: reading and writing, of course, but also arithmetic and the method of science. Any subject matter or theme must be an occasion to practice the basic skills, yet subject matter cannot be alien to daily lives and immediate interests. The teaching style is to start form the students' knowledge and understanding of things: how they view family and friends, school, work and leisure, health and disease, etc.

The first stages of literacy and mathematics are concrete enough, structured in themselves and challenging to motivate work which starts from the student's familiar realities. But for other basic skills, for example, those necessary to obtain information from the natural and the social environments, those necessary to process such information and to draw conclusions, those necessary for decision making (i.e., the skills involved in the methods of both the natural and the social sciences), it is difficult to decide which subject matter is relevant to the student beyond his or her daily experience. The inner logical structure of each branch of knowledge is not enough to challenge students of this age in these particular social and economical conditions to work seriously for prolonged periods of time; nor a somewhat arbitrary decision on the part of the teacher about what is "important", be it themes or even skills in a particular field.

What seems to be the answer to the problem of motivation and functional mastering of basic skills is to engage students in real actions relevant to them, to their peers and to adults in the community. This is why Project 11-14 has promoted productive or service activities on basic needs which challenge both the local population and the public agencies officially responsible to help satisfying them. Two such activities are beginning to meet the challenge, the first one being the recovery of rain water and the recycling of sewage water (and its further use in school and family vegetable gardens), the second being the care of children under five years of age. Although the activities are thought of as means to help the students, they are not only school activities, but real community actions; the responsibilities rest not only on teachers and students but on experts, public officials and local people. What I am reporting here is about the care of children under five as a community action in which students work as a part of the science education component of Project 11-14 curriculum.

The approach to under fives health care was developed in West Africa by Morley (1973), and adapted succesfully to an urban area in the city of Chihuahua, the capital city of the State of Chihuahua, Mexico (Centro de Orientacion Materno Infantil, et al., 1979). Through these programs, mothers are provided with means to participate in the health care of their children. The mean is very simple: a card called the "Road to Health" kept by the mother and in which the weight development of the child is recorded. Also recorded on the card are all of the vaccinations and other data relevant to the health of the child (order of birth, reasons for special care, nutritional supplements, etc.). Mothers are persuaded to take their children every month to a small clinic to be weighed; vaccinations are administered when indicated. The visits are occasions for the nurse (or the doctor) to talk with the mothers both formally and informally about health matters. This approach to health has made it possible to vaccinate children with an efficiency rate higher than 85 % and reduce infant mortality rates by 70 % as reported by Morley (1973) and the Centro de Estudios Generales (1979). Other benefits are less tangible although by no means less important: a healthier general development of the children, prevention of disease for the entire family, lesser expenditures in medicines and hospitals, more control of one's life, and so on.

The equipment necessary to carry out the program is found in every neighborhood clinic: scales, refrigerator to keep vaccines, first aid instruments, a few common drugs, contraceptives, etc.

The program relies on the cooperation of the local clinic and is coordinated by a full-time nurse and a group of promotors (one for each Project 11-14 educational Center). In each Center students who accept to work from two to three hours every day after school are given scholarships (about 20.00. US Dlls per month). The first activity consists of the students taking a census of all children under five in the community and registering their development by means of the Sakir band, which measures the arm's circumference and is an indication of the state of development of the child. The census also registers the vaccinations received by each child and offers the base to calculate the rate of infant mortality in the community. Once the census has been taken, students are assigned to specific families in order to remind mothers that the time has come either for the monthly weighing or for some vaccination. The students also visit the families periodically in order to register important events in the development of the child, like the time he or she begins to sit or walk. Students are also in charge of providing mothers with full fat soy flour when prescribed by the clinic. This flour is prepared by students in the school following a very simple method.

In the classroom component of the curriculum, students and teachers explore what is most immediate to them: nutrition for energy and growth. Once they have explored the issue, data about their community or similar communities are presented in order for them to realize the timely need of caring for children under five, many of them their brothers and sisters.

Next the full project is presented as a feasible way to participate. From then on, the school program branches off to study, through activities, as many aspects as can be seen, all related to nutrition, health and disease: disease as the result of deficient nutrition; infectious diseases and their prevention; accidents and intoxications; vaccines, their history and function; the main nutrients in foods, including local foods; the origin and distribution of the food they consume; prices in the markets and why they change; monopolies and advertising; etc. At the same time, specific knowledge is required from the methods of social sciences and mathematics to handle the census, samples, interviews, questionaries, averages, graphs, percentages, tabulation of data, balancing of diets, etc.

In this way, knowledge that otherwise would seem dispersed and so unlikely to motivate a student, is seen integrated and functionally connected with an activity in the neighborhood which presents action and knowledge both as necessary means for the well-being and development of local people. At the same time, every opportunity is seized to practice the basic skills which should constitute the specific purpose of elementary public education.

Project 11-14 contemplates a curriculum to be completed in two to three years. The first year (called level I) is devoted mainly to reading and writing skills, with growing math and science components from the first quarter to the third one. The under fives health program is the third quarter of this level I; students working in this program learned reading and writing only three months before.

References

Centro de Orientación Materno Infantil, et al., 1979. Experimental Program in Child Health Care, Cerro de la Cruz, Chihuahua, México (mimeographed).

MORLEY, D., 1973. Paediatric Priorities in the Developing World, Butterworths, London.

NETHERLANDS

BIOLOGICAL EDUCATION AND COMMUNITY DEVELOPMENT IN THE NETHERLANDS
A SURVEY WITH PARTICULAR EMPHASIS ON THE ENVIRONMENT

by L. Vlijm

1. Introduction

In the Netherlands the same situation was found as in other Western-European countries: that is that biology is only recently known under that name. About fifty years ago only botany and zoology were known, and together they were called "Natural History".

By this word the contents of the curriculum, at universities and secondary schools in that time is very well indicated: though a relationship existed with science (Mathematics, Physics and Chemistry), biology was almost totally descriptive, with some exceptions in the field of plant and animal physiology, which, however, had only just made a start.

The interest for plants and animals dated back to the "Cabinets of rarity" in the 17th and 18th century, which were small museums in which "rare" plants and animals (mostly of other continents) were collected and demonstrated.

This made that people became interested in the whole of living organisms in the world. The teaching system was closely connected with the presence of botanical gardens and herbaria at on hand, zoological gardens and musea at the other hand. Therefore attention was mainly given to systematics, morphology and anatomy: the descriptive branch of biology.

In recent times, especially after the new systematics in 1930, and the development of scientific technology, the experimental approach was started. And it must be said that at that time it seemed that only the experimental approach was biology and could be appreciated as a "hard" discipline. The system of Popper was applied to biology, and other approaches more or less neglected, or appreciated as "old-fashioned".

A new dimension was given to biology by the development of molecular biology and environmental biology. They were first introduced at the universities, but afterwards found their way back into textbooks for the secondary schools. By the specialisation, which occurred during the last 25 years in biology, the danger existed the field was desintegrated, and it must be said that this danger has not been followed by immediate actions of biologists to counter it.

As happened in earlier periods, when medical fields as well as agriculture were set apart, and were organized in other schools, it happens now that some of the biological fields are part of departments of chemistry and physics (molecular biology), or of environmental schools (environmental biology), in which only parts of biology are included, and others neglected.

Though it must be stressed that biology should be presented as one discipline, especially at the secondary school level, to meet the demands of the community, we will discuss in this paper especially the environmental biology in the Netherlands, with special emphasis on the secondary school level (the entrance to those who will afterwards have more directing positions in the community) and the community at large.

2. Short History

Biologists knew much about life, even in earlier days (probably some of them knew quite a bit more in a large field than present biologists can claim!), but they lived in ivory towers. They collected their plants and animals, gave them names and studied their relationships but they were known as people with a container on their back, collecting their plants (nowadays they do it with plastic bags), or with a net to collect butterflies or other creatures (nowadays again in polythene tubes).

Even now people will easily determine somebody with a green jacket, rubber boots, field glasses and, when male, with a beard, as being a "biologist": a rather common bird in the field.

I think that, in the Netherlands, an important move in the direction of making people aware that their exists an environment with living beings other than humans, was given by two school-teachers, Heimans and Thijsse who published, in the beginning of the century, several general, and even now much appreciated, books on plant and animal life in different habitats (e.g. dunes, heathes, ditches and lakes, etc.). Cooperating with them, a biscuit-industry published in the twenties and thirties albums about different habitats, with a good text, and plates, which again nowadays are highly appreciated.

Both of them, with others, contributed to the founding of the Society for the Protection of Nature (which actually came into being because the town of Amsterdam had decided to use one of the lakes in the Western part of the country as a rubbish-dump; nowadays it is the oldest nature-reserve in the Netherlands, because it was bought by the Society for the Protection of Nature). This organisation did a lot to come to the conservation for the future of the small "nature" part of the Netherlands: only about 7-8 % of the area is more or less natural, though influenced by man.

The last few years another society has been founded, which is of the "action-type". It is called: "Foundation for Nature and Environment", and publishes a journal, once every three months, indicating problems of planning and coordinating action of groups: in short against further demolition of the environment.

Special attention is given to the area of the Dutch Wadden Sea, which is very important for migrant birds (the population of some species is almost totally concentrated in this area during winter time) as well as feeding grounds for young fish of the North-Sea. This organisation, which operates on a private base, is very active.

In general it should be said that the impact of biology (by education) on the community is rather restricted: it appears that only a small percentage of the population gives priority to the environment. Most of them are more interested in salaries and social wellfare.

3. Problems

One could wonder why those who are aware of the importance of the environment are so few in the Netherlands, one of the most developed countries in Western Europe, with very high life standards.

We shall try to analyse these problems, which clearly are important when one tries to influence the community by biological education.

Some of them probably are rather general, and others will apply especially to the Dutch situation.

a. Panem et circenses. The Roman writer Julianus said in his

"Satirae":"Panem et circenses vulgus Romanum semper optabat": "The people of Rome have always asked for bread and plays". This indication that the population of Rome had only interest for food and in amusement in their leisure time, and thus had no interest to develop themselves, surely applies for the present situation of the majority of the world population. They are not interested in biology, and they are not interested in nature and environment. They are interested in food, in money, in using the environment so that they might have a "better life".

They are only few which, c.q. because they have animals as pets, or because they love flowers, will be interested to increase their knowledge about nature and to understand how it works, and how the impact of mankind on the environment might be mitigated.

When I am right this is one of the most serious problems to overcome when we are discussing biological education which is community-directed. I fear that this does not only apply for the Dutch situation.

However there are some particular characters of the Dutch situation which are of importance. It might be that some of them can be transported to the situation in other countries.

b. The man-made Dutch landscape. It is well known that the western part of the Netherlands is below sealevel and is only protected by dune ridges, dikes, and an elaborate system of sluices from being covered by the sea. (It might be remembered that in 1953, more then 2000 people were killed during extremely high floodings, which inundated the South-Western part of the country). The landscape in the Western-part of the Netherlands is manmade. There is a verb: "God created the earth, but the Dutch created the Netherlands".

This, to my opinion, is important for understanding the Dutch attitude. Within the government the Department of Watercontrol is one of the mightiest departments: originally it was a state within a state.

Being pioneers, and claiming land from the sea, protecting this land afterwards, the Dutch feel themselves creators, and thus have no deep feelings of respect for nature. They are managing nature and have been doing so for centuries and centuries.

So the attitude is prevailing that "nature" exists to be used by human people.

c. Religious elements. The northern part of the Netherlands was predominantly orthodox-protestant, and the South Roman-Catholic. Though the religion nowadays does not play such an important role as it did before (Protestants and Roman-Catholics nowadays are working together e.g. in one political party: C.D.A), it is evident that the relation between man and nature historically was influenced by religion.

The attitude of the christian religions in principle is given with Genesis 9:2 in which it is said that man shall be feared by all living creatures and that all of them are given in his hands. This seems not to indicate that man has to be kind to all living creatures: he can just use them, being the "crown of the creation". Even in this century the

Pope Pius X said that a christian had no ethical duties to any animal: therefore he was rather sceptical against the conservation of nature and the protection of animals. It should however be admitted that the spirit of the Holy Scripture indicates feelings of well-behaviour, also towards animals. The way in which the Christian Community in the Netherlands was taught to look at nature and living things, was influenced by the idea that they just were part of the material world, and that any relation with the spirit of God was unendlessly more important. From the side of the churches no attention was paid to the attitude of humans towards other living beings in nature.

d. Population density. It is known that the Netherlands, with at present a population of about 14 million people, are one of the Western-European countries in which the density of the population is highest. Of these population about 10 million people are concentrated in the Western part of the country, which is about 1/7 of the total area. This part is called "Randstad Holland", and the major part of the population is concentrated between Amsterdam and Rotterdam.

This area has only little natural reserves left. The major possibility for recreation are found at the beaches and in the dunes, which are used at the same time as water-catch areas and water-winning areas. (It should be remembered that water in the Netherlands is rather scarce: the total volume of water being 5 km^3, to be compared with Lake Como, with 27 km^3 and the American Lakes with 25.500 km^3.)

Because of the high density there is a need for large housing areas, and much space is taken by highways connecting the centres. As industries and governmental organizations are also crowded in the Western part, there is a need for recreation areas. However the Western part of the country mainly consists of polders, with controlled waterlevel, with a silt soil, normally used for cattle-breeding and thus not very easy to be used as recreation areas. They are good for cycling and angling, but do not serve the general public. Therefore man-made parks were and are constructed in these polders, especially in the neighbourhood of towns. As an example we refer to the "Amsterdamse Bos" an area of several hundreds of hectares. During weekends more than 100.000 people are recreating.

This indicates again that Dutchmen are managing their environment for a great deal, and that they have no high appreciation of natural sites, because they think they can be "created". Furthermore it should be noticed that only 7-8 % of the total surface of the Netherlands is more or less "natural", though even these parts have been under continuous influence of man.

e. Schoolteaching. Another thing to be mentioned (probably the same will be found in nearly all countries) is that, though in recent times there has been an amelioration of school programs in biology, thanks to an upgrading of the level in teacher-training colleges, even now the knowledge which is taught sometimes is not very stimulating. One could wonder if knowledge about the sexual reproduction of ferns and mosses can be used by a teacher at the primary school level. Children will probably not be much interested.

4. The actual situation. The impact of biological education on the community of the Netherlands is not easy to describe. We will only indicate the media by which biological knowledge is conducted to the society, and, the other way around, by which parts of the society have access to biological knowledge.

Societies

Of course in the community of the Netherlands a vast system of all kinds of private societies is found. Among these quite a number are biology-directed. Owners of cats, dogs, aquariumfishes, birds or other pets frequently are members of a society which organizes not only shows but sometimes has also some more or less frequent leaflet.

It must, however, be doubted if effectively biological knowledge is dispersed in the community. However, somebody who has a pet-animal probably will not immediately destroy all other types of animals.

The same applies of course for plant societies. Though the members of these types of societies sometimes have a lively contact, the impact on the community as a whole is small.

Action groups. During the last 10 years (it started in Amsterdam with groups as Provo and Kabouter) actiongroups have come into being. They should be seen as some sort of project-directed, and they try, with all means, to achieve their goals. They are not allways biology-directed, and some nostalgic elements most of the time can be detected.

They are only partly appreciated by other people, and probably never by local authorities. Their "no" is very forceful, but the production of alternatives is rare. Their impact in living quarters sometimes is high, but they are not organized groups and as such their influence on the community as a whole is small.

Nature Societies

We do not have to pay much attention to those societies who have nature-protection of nature studies as a direct aim. They are found in the same way in other countries.

It is true that they are important - but they are only for the happy few members and their influence on the community again is small.

Perhaps we have forgotten some groups. That is not so important. To my opinion we can say that, in general, two types of activities which could be of importance for the community may be found:

1) Those who love nature, or special living beings (pets or plants). They

know a lot about nature, sometimes they do only enjoy it for reasons for beauty. They are not necessarily interested in the intricate web of life. The environment is the environment as they see it: they are not interested in concepts, and probably they will not read a book to increase their knowledge.

2) Those who are the idealists. Nature is part of the universe and man therefore is interwoven with it.
   They are not interested in biological phenomena as such, though they will use the knowledge of biologists to reach their goals. Sometimes they are a danger for the biologists, because, on the basis of lack of knowledge, they either misinterpret data, or use only part of it, so that there arguments no longer are logic.

Education out of School
We will only shortly discuss the other ways to educate people in biology for their own well being.

Media as radio and T.V. sometimes communicate very well biological knowledge. They reach those who are already interested, and it must be said that sometimes especially with T.V., the comments are rather poor: nature-films are presented for entertainment, more than for education.

Another means of presenting biological data is through small shops, or small exhibitions in or nearby nature areas, or recreation areas. They are run by amateurs who like to contribute to "do" something for nature. Though they reach quite a number of people, this is especially so during the one rain-shower on a beautiful day!

Mostly, even in these activities, those are reached which already know. The not-knowing are not interested and they form the majority of the population.

Education in schools. Ministers of education (we have had 3 in the last ten years) do not fulfill the testaments of their predecessors: each has his own more or less brilliant ideas, not only about the money and the division of it among different types of schools, but also more philosophical ideas e.g. how each child, independent on the salary of its parents, should find their way in the education-maze: the children do not yet behave as trained rats!

This means that time-tables and curricula are in a continuous change, especially as far as secondary schools and universities are concerned. It must be really doubted if there exists nowadays one person who knows how the overall-structure is. In biology there is the danger that it will be included in "science" at the secondary school-level, and that science as a whole will be reduced, to make time for "sociology" (this term is used at large). The contents of courses are, as they were in the past, in the earlier classes very much dependent on the activity and interest of the teacher. I fear that most of them will be more stimulated to teach those who are already interested, than those in which interest has to be evoked. Furthermore the teachers are confronted with young people who are not very eager to be taught, because they fear that no jobs will be available afterwards. Also they are soon tired and more inclined to watch T.V. or have a party, than to study "dull things", of which the importance is not clear at the first sight.

The secondary school system is very much in the move, and it is not easy to predict what will come out.

It is very necessary that biologists clear their own house, and this applies for secondary schools as well as for universities (An ECBA workshop has been dedicated to the biology-curricula at the secondary level early 1980, which was to ensure that the situation of several countries will be discussed and general ideas can be developed).

It is necessary, when we discuss biological education in relation to the needs of the community that we have at least some ideas about what is essential. We are sure that we cannot give a complete picture, but we shall try to give some personal reflections. They should, together, give an idea about the aims of the education system in biology.

The Aims

1) The concept diversity, and mutual relationships. For a biologist diversity is a simple concept as such: we know that there are more than 350.000 higher plant species, and several millions of animal species. It is even tried to measure the quality of an ecosystem, or community, by measuring diversity. Indicatively this is sometimes called: species-richness.

For a community the first thing of the valuability of keeping diversity is that otherwise life will be monotonous. It is easy to make children aware that some streets are duller than others, mainly because of the monotony of the houseblocks, and in the same way when, in a town, only one species of tree is planted, a dull city appears. However this does not mean that continuity should not be there.

Mutual interrelationships are in the same way important: the feeling that any specimen of a species who is alone is a poor one, can easily be taught to young children. It is a notion which can be widened (social contacts, as well as food-relationships).

In general it may be said that, when young people can be made aware of diversity and of mutual interrelations - not in the field of knowledge, but as an understanding of reality - they might be able to detect the same thing in human communities.

One could wonder if the same concepts could not be demonstrated more directly, by using the human society as an example. This leads to two types of questions:

a. Is knowledge about nature important for the knowledge of humans?
b. Will biosociology be accepted?

We can not discuss these questions now, but they are important to be covered.

2) Understanding of the nature of scientific research: its open activity. The "accepted knowledge" of the present is changing. This is difficult material. It will not be easy to introduce this to children. On the other hand the "grasp" of it will be important. It needs reflection to come to a practical approach.

One of the dangers it includes is that young people are easier told something unflexible, than something flexible. Thinking in nuances seems easier for grown-up people, on the basis of their own experience: their unflexible attitudes have been worn out.

3) The limitations of the scientific (in this case = rationalistic) thinking. Many problems, even in the field of biology, cannot be tackled alone by scientific methods. In ecology the concept ecosystem exists (cf. Schaefer, 1979). It can be analysed and leads to inclusive thinking. The way of preservation is logical and follows from scientific analysis. On the other hand a historical component exists in the ecosystem. Its development through time is not known. The intricate web of interactions of prey and predations, of plants and herbivores can be described, but most of the time they cannot be studied through time. Accidents, happenings that occur by chance, are probably very important, but they are neglected in the scientific approach.

It is evident that in human society some are more "lucky" than others. Some are always in the sun, and other walk in storms and rainshowers.

It is of the utmost importance that people realize that there is not a comprehendable linear causality, which determines anybody's position in life. There is more. To my opinion this "idea" can be brought forward in the teaching. It needs, evidently, more attention.

4) Understanding of the biological basis present in fields as medicine, health service, agriculture and nature protection. It is evident that this is a more nearby goal. It is evident that this should, in the list, not be the first goal, because we are not defending ourselves, or our position, but try to give people something which leads them to an insight into the relationships between different fields of knowledge. Persons should not be cupboards in which knowledge and insight is collected in separate drawers.

5) Appreciation for the beauty and other emotional feelings evoked from phenomena in nature.

As already was indicated living creatures cannot only be understood through science by scientific methods. Artists also have their relations with nature, and, in fact, in every person several aspects of appreciating living nature are found.

It is evident that, when respect for the living world is present in a child, the grown-up will be open to it, and better chances exist that it will not be destroyed.

6) Understanding of the development of concepts in biology, and their dependence upon the social structure of the community, as well as the technological status of this community.
   When this can be taught, it is clear that knowledge is promoted, which enables people to accept that even in the present time differences in the attitude of communities towards nature exists.
   This is very important. People, now demonstrating with Green Peace against the killing of seals should be aware of the fact that their forefathers killed seals and whales in the Northern Seas, and that even their fathers may have been among those who killed the last blue whales in the Antarctic.
   This is not to waken the sake for the protection of wild animals and plants, but to accept that the attitude of some present communities may be the same as it was in another country in the past.

7) An understanding of part of the present knowledge concerning basical biological problems e.g. evolution, development of the embryo and genetics.
   Evolution clearly is a process that should be taught by biologists. On the other hand embryology and genetics, both of great importance for any community are based on the research of biologists, or at least the research is done in close cooperation with biologists. Furthermore in animals, sometimes, more is known about the principles, and therefore it is easier for a biologist to cover the field than for somebody who concentrated mainly on man.

8) The most important, because it covers quite a number of the other items, is a basic appreciation of the position of man in nature. The knowledge that man himself is a living creature among others, and that he, by his dominance influences all other living beings, is prerogative.

The Future

I am not in the position to develop major ideas how the situation might be changed.

I think, however, that, in general courses for everybody attention should concentrate on man, and give data on

a. man's position - history (evolution)
- present (environment)
- future (population-control)

b. man as an organism
- body structure - functions
- health and hygiene
- nutrition

c. Man among others
- sexes
- social contacts.

Who can write the booklet for teaching children of the primary school?

Some Conclusions and Remarks

1. Biological education is not (or only rarely) open to the demands of the community and the community is not asking for biological education (Questions: What is "the community"? What is "biological education"?)
2. Partly this will be caused by historical and culture (religion included) reasons, but evidently it depends also on the fact that biologists do not "sell" their knowledge.
(See: "Biologists in European Society", ECBA 1979.)
(Questions: What will biologists sell, to whom, how and why?)
3. It is very important to make children aware of their environment.
(What is environment for a community?)
4. Start with primary schools: otherwise it will be too late.
(See: suggestions of items at the primary school level)
5. It is impossible to reach all pupils.
(How reach all pupils?)
6. Grown-people will appreciate their environment,
a. because they like it
b. because they are punished with mis-behaving
(What is the best strategy: Punishment or Attraction?)

PHILIPPINES

BIOLOGICAL EDUCATION AND DEVELOPMENT IN RURAL COMMUNITIES OF A DEVELOPING COUNTRY

by Dolores Hernandez

Introduction

In 1977 the estimated population of the Philippines was 44,4 million. With a growth rate estimated at 2,8 %, the population at this time would be about 47 million. Of this number about 32 % live in urban areas. Hence, any movement intended to bring about changes in society and the life of the people must consider the rural communities where about 68 % of the population live.

These are the physical realities. There are also the socioeconomic realities. The current gross domestic product per capita is US $ 300 in the island of Luzon (the largest island) and less in the other regions of the country. In the next 25 years the gross domestic product is expected to double every 4 years. But much of the growth will be taking place in the large southern island of Mindanao where the per capita income is presently about the lowest in the country. In introducing innovations therefore, it is important that these are within the financial capabilities of most people in the community.

The preoccupation of the leadership of all sectors of the country, education included, is presently focused on improving the quality of life of the people.

This preoccupation is not confined to the social sciences. The Minister of our National Science Development Board in a recent speech on policy pointed to the fact that the thrust of the government is countryside development. ... "The science and technology system, is, therefore, expected to focus its attention to and orient its plans and programs towards the development of the countryside." (Magno, 1977)

Nor, is this thrust peculiar to the Philippines. At a recent meeting of ministers of education and economic planners of Asia and Oceania, one of the declaration of the conference stated "it is clear that education, whose ultimate goal is to promote the full development of the human personality, will be faced with increased responsibilities for contributing to the economic, social, cultural programs of societies..." (Final Report, Tenth Regional Conference, 1978).

Development therefore in the Philippines as well as in many Asian countries is looked upon as having strong economic and social-cultural thrusts. This paper deals with the use of science education (with emphasis on biology) in meeting the problems of development as this affects the larger proportion of our people - the 68 % who live in rural communities.

Current Efforts to Determine Skills Needed for Rural Development

To implement the goals of development, the Philippines is presently conducting national programs designed to improve the life of the people. Some specific projects include: green revolution, nutrition education, population education, youth and community development, cultural renewal, tax consciousness, drug education, agrarian reform, sports development, physical fitness, and beautification campaigns.

In connection with all these projects, there is a demand for the direct participation of the schools. These demands imply that teachers have to be equipped with the skills and competencies of their new roles and tasks.

As a consequence, a research unit of the Ministry of Education and Culture conducted a study on the tasks and skills required of teachers for effective participation in community development projects. Rural communities were classified as agricultural, fishing, industrial (small scale), and fishing - agro-industrial communities. This study reveals that the activities the teachers perform do not differ from one type of community to another; the skills needed to perform these activities were identified as: (1) functional knowledge and practical skills in the activities concerned; (2) human and public relations; (3) proficiency in communication; (4) group dynamics and management; (5) gathering and interpretation of data and (6) evaluation.

It is true that the types of teachers' activities may not differ among the different types of communities if we take an overall and general view; but anyone who has done a community-based project will realize that the specific activities differ from one type of community to another and even from one community to another of the same type. This is so because meaningful activities designed to improve the quality of life are community-specific and people-specific. By this I mean that the problems of development are very closely related to the specific environment, its resources and constraints. It also means that unless the people are personally interested in any given activity, it will not take root.

We should note that the study just mentioned focuses on professional skills that teachers need in order to implement rural transformation projects. It does not touch on the content of the possible projects a teacher may be involved in, nor does it deal with provision of specific materials that teachers may use. In this connection it should also be pointed out that colleges of education are concerned with training teachers for teaching specifically in rural communities. In fact, a number of them are introducing innovations in training teachers for rural areas. But here again the focus is on acquiring professional skills.

One more point should be brought out, at the village level, the teacher has to assume more leadership roles much more so then her counterparts in the larger towns and in the cities. Furthermore, the village teacher has less access to sources of information or to consultants and specialists who could help meet certain village problems which may require more technology/knowledge than found in the village.

Thus, although she may have the professional skills, the village teacher is handicapped by a lack of substantive knowledge which can be applied to help her cope with rural development problems.

Implementing Education for Rural Development

On a national level there is a major project which focuses on education for rural development. This is known as the Youth Civic Action Program (YCAP). It involves all students of elementary, secondary and higher education institutions. Every teacher of a subject is responsible for implementing the program in connection with the subject she teaches. Lessons may be either structured around community problems or they may culminate in civic action projects which contribute to community development.

For college and university students a Youth Summer Civic Action Program has been developed. Students spend 5 weeks of one summer, during the third year of their course program, in a rural community to work on a development project related to their major field of study. This is a requirement for graduation.

It should also be pointed out that all heads of government offices and others involved in supervisory work must spend 2 weeks of each year working in a rural area. Arrangements for all these activities for both college students and government officials must be done through the local government officials who assign these volunteers to priority projects of their communities. Some choice in the specific type of activity or project is possible provided this falls within the priority projects of the community in which the service is to be rendered. In brief there is an attempt to match what the individual feels he can do and what the community needs.

This is the organizational scheme on the national level; there are however many projects with similar goals of improving rural communities which are being undertaken on small scale by various institutions and agencies. For example, a survey made in 1975 revealed that there were 138 curriculum development projects in the country dealing with the following topics: agricultural education, drug education, foods and nutrition, health education, population education, trades and work education. (Sumagaysay, 1975). These were mainly "enrichment" materials rather than textbooks. It can be safely assumed that these materials deal more specifically with local needs or problems.

Rural Transformation and Village Technology: Some Characteristics and Procedures

The projects of the Science Education Center in rural development fall into the category of small community-based projects. It became obvious right at the start of Center's involvement that to help a community improve itself, several aspects of life need simultaneous attention. Thus, an integrated approach was used, rather than one based on purely one discipline such as biology. However, the staff constantly kept in mind the possible implications of these experiences to science curriculum development, for this is the major aim of our science education efforts.

In describing the characteristics of our rural education projects, I would like to begin with a concept of village technology which was defined by a meeting held in Nairobi in the mid 1970's during a UNICEF seminar on simple technology for the rural family. Their definition follows.

> The essence of village technology is technological simplicity on the small scale that is needed in rural areas. Village technology does not seek to bring about an immediate technological revolution in rural areas. Rather it seeks to introduce small initial changes which are practical, easily possible and, which when taken together have the potential to bring about significant improvements in the overall quality of rural life, thereby providing a basis for more advanced development. (UNICEF News, Issue 90, 1976).

Some characteristics of our projects are incorporated in the foregoing definition: small scale, practical, easily possible, designed to improve the quality of life. Other characteristics and procedures through which they were achieved are described in the paragraphs that follow.

1. Determination of specific needs at the village level. This implies acceptance of the project by the community, involving the people in

responsible participation as early as possible, conducting a survey of the community, informing the people about progress of the project at all stages. Some of the information which may be gathered about the community include economic activities; resources of the community (human and material, used and unused); constraints in the community, e.g., accessibility; beliefs and practises associated with their occupation, daily life, natural phenomena, health and nutrition; social-cultural perspectives; knowledge about their environment; skills the people feel they need to learn to improve themselves; existing indigenous technology; civic activities and responsibilities. In most of our projects, the school administrators and teachers and the local leaders have been used as entry points into the community.

2. Analysis of the data. From the data of the survey, problems can be identified and implications for the curriculum and nonformal education can be made. A simple way of analyzing data can be used as shown in Table 1. In the first column, data from the survey are listed. Next, problems related to the data are determined. The problems are based on one's knowledge about the community and may also be inferred from the data. Next, implications for curriculum development are determined. What knowledge or skill aspect of the problem may be introduced in a learning situation? These are listed in a third column. Finally, implications for nonformal education are listed. Once such a table is made, plans for the various activities of the project can easily be prepared.

3. Specifying the areas for curriculum development and determining points of entry. Our experience with our own projects have shown that teachers tend to be conservative and to follow the school syllabus or textbook closely unless assured by their administrators that they can introduce new relevant materials. They are also more likely to do so, if the materials are provided and made available.

The following broad areas have been identified in several of our projects.

A. Health, personal and family
- pre-natal care
- infant care
- nutrition (based on local foods mainly)
- dental care
- first aid, use of medicinal plants
- sanitary toilet habits

B. Sanitation of the environment
- garbage disposal
- drainage
- eradicating breeding places of flies and mosquitoes
- ensuring safe water supply

C. Environmental pollution
- water pollution (including the sea and beaches)
- pollution caused by man and animals
- air pollution (as in charcoal making)

D. Community welfare
providing water supply
purification of drinking water
conservation problems as: deforestation, overfishing

E. Economic welfare
need for cooperatives among farmers, fishermen
alternative economic activities for seasons of low productivity
improving indigenous technology: e.g., better ways of making charcoal, making better fish nets, constructing fish pens, using forest products, etc.
suggesting alternative economic activities
improving food production e.g., scientific farming, utilizing coves and inland water for raising fish

F. Food preservation
improving indigenous ways of food preservation
"borrowing" successful ways of doing things from other communities

G. Cultural welfare
providing reading materials
developing reading materials - the community's natural and social environment

4. The project activities in which the adults participate:
deal with problems perceived by them
benefit (economically and in other ways) most if not all the participants
utilize materials and labor available in the community itself or in nearby towns
are self-propelling, i.e., once learned they could be carried on by the villagers without any supervision
do not rely heavily on external resource people except during the initial stages
encourage development of self-reliance
consider constraints in the environment, and
consider the human response to the introduced technology.

5. Evaluation of the project is done periodically. This takes the form of: visits (ocular inspection) by the project team to different parts of the community; interaction with the teachers and with parents; school visits. In our project in the fishing village we gave a test which dealt mainly with health, sanitation and knowledge of communicable disease prevention. The adults as a group did better than the children at grades III, IV and V levels. But the children at grade VI level did better as a group than the adults.

6. The products of the projects take several forms such as an improved technique or a useful product. But all the projects have each produced some modules written in either English of Pilipino, and intended for formal and nonformal education. Such modules become a

major way whereby knowledge or skills gained by the original group of project participants are shared with the rest of the community. The modules may also be used in other communities which have similar problems and environment.

7. Lastly, our experience in various rural science education projects provides us with added insights into the process of development. It is closely associated with the learning process of a community or a society as the individuals that compose the community move towards achievement of specific project goals.

Projects at the Tertiary Level

Many of the current projects both of the UPSEC and other institutions involve the elementary and secondary levels of the formal system education, and the corresponding age groups for nonformal education. The tertiary level has been much less involved, although the machinery for their involvement exists, e.g., the Summer YCAP for college students, and the rural service required of heads of government offices and supervisors. This is not to say however that the tertiary level is not interested. A recent publication listed and described 35 of the country's outstanding projects in community transformation through nonformal education (NRDCTE, 1977). Sixteen of these 35 projects were initiated and are being conducted by institutions of higher learning. However, the stress of these projects is on nonformal education. It is also interesting to note what a group of graduate biology students recently listed as the major problems of their communities to which biology topics could be related (see Section IV for this list). These students have not been involved in any of UPSEC projects, but they came up with similar broad areas which many of our community projects have dealt with.

A group of university people organized an interdisciplinary and interagency workshop to discuss major problems in this country and to suggest some technological or institutional solutions. It will be noted that the first 2 problems (overcrowding, decreasing availability of areas for agricultural use and forest cover) cited in their list coincides with some of the problems listed by the students. See also Section IV for the list of problems cited by the organizers of the interagency group discussions.

The foregoing examples are cited to point out that there is awareness of the problem and people are doing something about using the disciplines to solve development problems of rural communities and society as a whole.

The science disciplines however at tertiary level have remained largely aloof from these movements. Involvement in rural improvement projects has been by individual; and the academic programs have remained largely unchanged.

What is needed is to provide time for the student in the biology program for some experiences which will help him see how he can use his knowledge in meeting some practical problems. He could also experience sharing what he knows with rural folk. Or, students could visit rural areas, live briefly there and then bring back to the University some community problems which could be worked on more intensively with help from other biologists.

SINGAPORE

BIOLOGICAL EDUCATION AND COMMUNITY DEVELOPMENT: SINGAPORE EXPERIENCE

by A.N. Rao

Introduction

The advances of biological knowledge have brought both prosperity and problems to many developing countries. It is through the biological education that communities or nations would progress further in the future years and this progress can only be achieved through a systematic and co-ordinated approach in biological education. It is an undeniable fact that we are living in an age when a large section of the human population lacks the basic human needs, and yet the scientist by his own brain power, research and technology has made considerable progress which have brought prosperity to certain nations and societies in the world. The knowledge that is presently available has to be well spread out and applied to bring forth certain positive changes to the developing world. Biosciences have to be utilized further to contribute and improve the life standards of the people in the developing countries and this can only be accomplished by a more equitable bioscience education and the proper application of the different scientific principles and techniques.

Most of the developing countries are scattered on either side of the equator in the tropical and subtropical belts with varying degrees of high temperature, rainfall and humidity. To what extent that these environmental factors have contributed to the lower standards of life in such countries is a debatable point. But, nevertheless, it is true that different pests, parasites, infectious diseases, high or low rainfall than normal, and floods cause very great damage and loss of human life and agricultural commodities, when compared with the temperate countries. Added to these there is the problem of over-population of human beings in the tropical countries, minimising the effects of advancements made through the application of science and technology. Examples of such countries are numerous and well known.

Curriculum Development

The question of deciding the quantitative and qualitative relationships and striking a balance between the topics concerning plants and animals that should be included in Biology teaching is always a difficult one. In our experience one of the overriding considerations should be with regard to the topics of local importance especially the ones that directly have an influence on the welfare of local people. For example cautioning the students about the dangers of soil erosion and deforestation is more important than teaching them about soil chemistry and photobiology of tropical forests. Priorities have to be established and certain aspects should receive top priority in teaching, including a proper analysis of the subject, the various problems involved and the possible solutions for them. The constraints of time factor and the number of teaching hours available would always impose certain limitations. The topics not covered at the undergraduate level would be further discussed adequately at the graduate level teaching so that all aspects concerned are fully covered and students are fully made aware of the regional and local biological problems before they leave the university.

While formulating the curriculum for Biology and its execution it has become increasingly clear to teach the students both fundamental similarities and differences between plants and animals. One should not replace one with the other and enough time should be devoted to structural and functional correlations in both plants and animals. In the final analysis it is the knowledge in these two areas that would help the man to choose the plants and animals for his own use and betterment. An understanding that there is in fact a difference in the philosophy of studying plants and animals can bring a better meeting of the minds in Biology. No one should confuse himself or herself by thinking that plants and animals can be studied with similar philosophies because of apparent overlapping of facts seen in areas like fine structure of cells, ATP metabolism, respiration cycles, cytology and genetics, clonal propagation and others. It needs to be emphasised that integration or synthesis of biological knowledge into workable topics or frameworks is suitable mostly for advanced courses and at the undergraduate level the topics of animal and plant biology should be treated separately. This approach has many advantages particularly in educating the teachers, administrators and curriculum specialists. It is only proper that biology should be under constant scrutiny and subject to constant criticism. Only then the biological ideas and facts be refined, clarified and applied for community development.

Utilization of Biological Knowledge for Community Development

Popularisation and use of Biology in Singapore most inevitably is in relation to man in the urban environment since there is no hinterland and the only resource available for further development consists of human beings. Therefore, no effort should be left unspared in providing a suitable environment for the comfortable living of the people. The proper Biology education is an important means towards this end.

Regarding the useful contributions: The local biologists have succeeded in increasing the food production, in spite of very limited land resources available, and by the use of hydroponic techniques. The application of modern methods of veterinary sciences has resulted in achieving self-sufficiency in poultry and pig production in the last few years and excess production has opened the outlet for exporting these products to the neighbouring countries. The production of aquarium and food fish is increasing and the former are exported in good quantity every year. Sea food production is increasing year by year by using floating cage net method in the coastal waters and allowing the fish, prawn, and mussel to grow in their natural but controlled environment. The straw and oyster mushrooms are cultivated in large scale by re-utilizing some of the agriculture waste. The cut orchid flower trade has exceeded the value of 1,1 million Singapore dollars per month (1 US $ = Singapore $ 2.1). The timber is cured and processed locally and made ready for export and the value of daily export is little more than 2 million Singapore dollars. The potentialities of biological resources, which are limited in Singapore, become manifest only when they have been brought out by the imagination and labour of the working biologist. The efficient and skilful trading would certainly enhance their economic value. In fact this is the operating principle of timber trade in Singapore. Creating an urban greenery has of late gained importance in the local scene. Anyone who has visited Singapore in recent years would appreciate the beautiful landscapes, avenue trees, parks and other recreation centres dense with planned growth of vegetation. The knowledgeable use of plants has helped to cleanse the polluted air, screen unsightly areas, deflect unpleasant loud noises and beautify the concrete and steel landscapes of the city. All the above mentioned biological activities have made a positive contribution towards community development and economic gains.

Demands on the Tropical Biologist

One of the greatest services that the tropical biologist can render to his or local community is to signal an urgent warning and to make them realize the value and heritage of the existing variety of living organisms in their respective countries and to stop completely their tragically accelerating decimation. The task of discovering, studying and classifying the innumerable species of the tropics is hardly well started in many areas. A thorough biological inventory in the tropics is yet to be completed, let alone a thorough organization of the needed information. These facts would caution us while establishing our priorities and using the biological knowledge for community development. Different kinds of animals and plants are not only being endangered but also becoming extinct. Often it is true that animals, because of their closer similarity to man in movement and reproduction, receive greater attention than plants.

In the tropics there is an ever increasing and urgent need to educate and train some bright biology graduates who can gather correct information from a vast data and coordinate the facts into a functional whole. If we stop exploring and generating the basic ideas in tropical biology, then the applied fields and their technology will decline in course of time. As it has been pointed out that the applied work is a "using" discipline rather than a "generating" one. These facts have to be well communicated to the concerned people by the working biologists. The call should be made to the leaders of different areas of activities, both in the private und public sectors and wherever necessary proper and suitable short courses in Biology should be offered covering all aspects of nature conservation problems. Popular talks with basic facts of some educational value should be delivered by the local biologists in the community centres, as are being done in Singapore. The use of mass media like radio and television for this purpose is common. With the German collaboration there is an educational television service established which show useful biological films of educational value.

Conclusion and Outlook

In conclusion may I submit that the tropical biologist has a major role to play in the community development of his country. He can discharge some of his responsibilities most effectively through education and training the manpower required to solve the various problems as we have done to certain extent in Singapore. The biology education in Singapore has been adopted to serve the needs of the community without sacrificing the scientific elements either qualitatively or quantitatively. Further improvement will no doubt be made in future to solve the problems that usually crop up from time to time. No particular solution can be satisfactory even to the same problem as the time passes by. New solutions have to be found out both for the old and new problems and it is in this area that the integration and application of pure and applied biology will become most rewarding towards community development.

ZAMBIA

BIOLOGICAL EDUCATION FOR COMMUNITY DEVELOPMENT IN ZAMBIA

by E.B. Rugumayo

1. Background Information

Zambia has a population of just over 5 million people, unevenly distributed over an area of 753.000 sq km, of these 5 million (40 %) live in urban areas, the majority of which are to be found along the line of rail which runs from south to north of the country, serving the rich and vital industrial Copperbelt with the necessary transport and other communication facilities.

On attainment of independence in 1964 Zambia found herself with a very poorly developed educational system, which was a legacy of colonial attempts to integrate Zambia into a white-dominated federation in which she would act not only as a reservoir of cheap human labour but as a producer of cheap raw materials, mainly copper, for export purposes. At the same time the country found herself with about one hundred graduates, and these were not trained in Zambia but outside the country for there were no institutions of higher learning, let alone a University.

For her economy, Zambia relied, and still does, mainly on copper, and for her food on expatriate commercial farmers (who employed/ employ capital intensive farming methods to the detriment of peasant farmers) as well as imported foodstuffs.

On the health front, the country places more emphasis on curative than preventive medicine; and possesses a comprehensive net work of dispensaries, clinics and hospitals, plus a medical school with a corresponding teaching hospital. Despite a massive injection of capital into a free health service for her population, Zambia still suffers from a high incidence of infant mortality, caused largely by malnutrition and gastro-intestinal disorders.

Peasant agriculture has been of a shifting kind, known as chitemene system of farming. This system has been aggravated by the commercial farming techniques which have left the peasant with no alternative but to depend on food from the city and to continue to rely on unprogressive methods of farming. It is important to note that the government has now banned the chitemene system.

With this rather sketchy background information, we will briefly summarize what Zambia has achieved in areas of education, agriculture, health, conservation, and how each one of these has a reciprocal bearing on the biology taught in schools; and what could be done to improve the situation in bringing about community awareness with regard to biological problems of the environment.

2. Conservation

This is an area of special concern for Zambia, because the climate of the country (4 months rainfall and 8 without rain) could give rise to scrub/desert conditions if careful husbandry of the resources was not encouraged and practised. Both the government and the public are beginning to awaken to the realities and potential gravity of the situation where conservation is not practised.

2.1. Zambia is a signatory to the 1968 Algiers African Convention on the Conservation of Nature and Natural Resources which was ratified by most DAU member states. The preliminary draft of this convention was prepared by the IUCN.

Article XIII - Conservation Education states, inter alia, that the Contracting States

> "shall ensure that their peoples appreciate their close dependence on natural resources, and that they understand the need, and rules for, the rational utilization of these resources," and "for this purpose they shall ensure that the principles indicated above are:
> (i) included in educational programmes at all levels
> (ii) form the object of information campaigns capable of

acquainting the public with and winning it over to, the idea of conservation."

In addition to the above, Zambia endorsed the Recommendations of the UNESCO Seminar on Ecological Principles for Development (Blantyre) Malawi, et al.) on environmental education. As a matter of interest the frame work of these recommendations was submitted by the Zambian delegation.

2.2. Conservation education at school level. Although some conservation education has been carried out among the population, nothing spectacular has been achieved at community level. For example, stamps to commemorate some important events in the conservation calendar have been issued, a Forestry Day and Conservation Day have been celebrated. However, it cannot be said that the adult population has developed an appreciable degree of awareness for conservation and its long-term-positive effects. Credit goes to the Wildlife Conservation Society of Zambia for its work in popularizing conservation. In addition to the above activities the society has published calendars, diaries, monographs, carried out advertisements and campaigns to popularize the importance of conservation.

Perhaps the society's singular and most spectacular achievement lies in its primary schools' CHONGOLOLO conservation programme. Chongololo means millipede; and both the programme and the society's magazine are named after this diplopodean crawler.

The society is known throughout the Zambian primary school system, and the concept of conservation is being planted at the grass-root level, and at an impressionable age.

The Chongololo magazine is sent free to all primary schools which have active Chongololo clubs. Twenty-fife copies of each issue (and they are issued quarterly) are sent to schools. Accompanying each issue is a Teacher's Guide. Some years back there were about 1.500 Chongololo clubs throughout the country, but now there are 600 active clubs with an estimates membership of over 20.000. A quick glance at a typical chongololo issue will show that the approach is both ecological/naturalistic and conservationist. For example, No. 3, 1978 issue starts off with what to look out for on a rocky habitant, in which a few of the major organisms are described. This is followed by a project on fire, then a write up on succulents, followed by an activity. The next pages discuss four friends who include the Lizard, Gecko, Skunk and Chameleon. Details of how to look after a chameleon are given, and finally club activities.

Club members are expected to keep Chongololo diaries, which should contain their observations and activities.

Any questions, problems and interesting observations encountered by the club members may be forwarded in writing to the societys' headquarters where a lively correspondence unit is kept busy answering questions. Some non-school adults have shown interest in the Chongololo programmes, and there are about 300 members outside of primary schools.

2.3. The Wildlife Conservation Society of Zambia has realistic and not so ambitious plans for the future. They would like to extend their activities to secondary schools and later to the public. This gradual and cautious approach is a result of meaningful experience the society has had over the years. Its funds are limited and suitable staff difficult to recruit. Given money and time, the society should be able to contribute effectively to conservation/environmental education in Zambia.

3. Nutrition and Health Education

In 1967 the Zambian parliament passed an act establishing the National Food and Nutrition Commission. Later in the same year, a governing council was formed, and office accommodation was provided for the commission. The funding came from the government, and grants from UN bodies such as WHO, FAO, UNICEF, as well funds from such well known organisations as Oxfam, Danida, and VSO were pumped into the commission's programme.

The Commission is charged with the responsibility of improving the quality of food and nutrition throughout the nation; and to advise the government and its relevant bodies as to what should be done to improve the food intake of the Zambian people.

3.1. Nutrition education in schools. Only two years after its establishment the commission launched a programme of nutrition in schools. A nutrition teaching Kit for primary school was produced in 1969, and about 4.000 Kits were distributed to all primary schools.

The Commission found it necessary to collaborate with the Ministries of Education and Rural Development in order to ensure that the broadest cross-section of both adults and those engaged in formal education was affected by the programme. Refresher courses on nutrition for Homecraft tutors for Teachers' Colleges are a regular feature of The Commission's programme. Supplementary feeding were introduced in primary schools and have been a regular feature until quite recently when lack of funds put a halt to this important item of The Commission's programme. School teachers are encouraged to actively participate in the Commission's activities and Teachers' Notes to accompany the radio programmes on nutrition are sent to schools free.

3.2. Nutrition education for adults. To popularize the work of the Commission, regular Seminars on nutrition are held for government officials, trade unionists and employers. The Commission's recommendation for provision of kitchens and canteens on work places is beginning to be implemented, but rather slowly.

As part of its campaign to popularize itself, the Commission has encouraged the setting up of food and nutrition committees throughout the provinces. These concern themselves with the feeding of adults and children, the growing of nutritive foods and keeping of domestic animals. Emphasis is placed on the scientific management of diets and the setting up of intensive nutrition units.

The Commission has field workers at various levels but the food and nutrition committees are run on a voluntary basis. Because of this, many of the original committees have ceased to exist, and the numbers of active ones decreased. This slackening has been aggravated by the serious economic crisis through which the country is passing.

Newspaper campaigns aimed at the wider public appear annually, radio programmes on nutrition aimed at adult audiences are heard in all the eight national languages, including English; a broadsheet entitled Nutrition News is distributed to nutrition groups (in various languages), schools, colleges, and health centres. "Nutrition songs" by a national popular singer have been produced on a record which is distributed free to the above groups.

The National food and nutrition survey has reported favourably on the impact of the commissions' programme on the feeding habits of the nation. Infant mortality has dropped, especially among members in the rural areas; malnutrition cases have also decreased. But there is still a long way to go.

The Commissions' immediate programme is to analyse and publish the results of the various surveys so as to enable it to structure its future programmes on the basis of meaningful data.

3.3. In conclusion, this brief report should not end without mention being made of a number of activities which are intended to educate the public about certain important aspects of biology. I have in mind a number of public seminars which are held throughout the year. These have covered such topics as pollution, alcoholism, drug addiction and high blood pressure. Each of these was preceeded by extensive press coverage, and public attendance was very encouraging.

Mention should also be made of the countrywide science clubs called JETS (Junior Engineers Technicians and Scientists) which are found in every secondary school, and a good percentage of primary schools. These hold annual shows at district, regional and national levels. Included in their projects one always finds biological investigations on conservation, health, nutrition, agriculture and genetics. These shows are very popular, and are always attended by a broad spectrum of the country's population.

Section III

Particular Issues

A SALMONID ENHANCEMENT PROGRAM AS A MOTIVATOR IN BIOLOGICAL EDUCATION

by F.A. Gornall

There are thousands of streams draining the water from the province of British Columbia into the Pacific Ocean. At the turn of the last century these streams nurtured annual catches between 300 and 360 million pounds of salmonids. Today the annual catch is only one half of that former harvest. There is no single reason for the decline. A growing population, modernization and expansion of industry, forestry, mining, hydro-electric projects and our vastly improved abilities to locate and catch fish have all contributed.

In 1977 a Salmonid Enhancement Program was launched as a joint effort by the Federal and Provincial Governments. Its goal is to double the level of present salmonid populations. Although the primary motive is to restore and maintain a natural resource at its previous productive level,the most far reaching benefit may come from its educational value as a motivator in the learning process. Interest in spawning salmon has always been high, so high that teachers will take their classes from Vancouver to Adams River, a round trip of 750 miles, to visit the Sockeye salmon spawning grounds. If this interest could be maintained, salmonids and their habitat could lead to a myriad of educational experiences that would cover most topics in biology and most biological issues of the community.

Salmonid Study in B.C. Schools

For many years the saga of the migrating salmon was treated at several levels in our school system under the heading of Social Studies. Salmon cycles were studied as - salmon cycles and their importance to the commercial fisherman. An emphasis was placed on the economic value of the annual harvest. As this value declined, much lamenting was heard but very little action was taken to reduce or prevent further decline.

Salmonid stories appeared only occasionally in literature assigned for reading. Science curricula barely recognized the existence of salmonids. Secondary school biology ignored them completely. Even after the implementation of the innovative secondary biology programs of the 1960's, which emphasized the study of the environment through an ecological approach, salmonids were still not included. Very little factual material was available for teachers or students.

Salmonids and a Course in Basic Biology

Instead of salmonids being examples quoted to support a few concepts in a traditional biology course, the educational process could be reversed so that the salmonids become the basis for a course in fundamental biology. It could be titled "The Salmonid Approach to Biology Teaching", and following the pattern of the B.S.C.S. biology series might be designated as the "Silver Version". All of the traditional divisions of biology would retain their identity and much of the descriptive material presently utilized would still be used, but subordinate and in comparison to the corresponding material applicable to the salmonids. A brief survey of the content of the silver version would show how these traditional subdivisions would fit together.

Classification and nomenclature: The eight species of salmonids would provide a framework to which all other organisms could be systematically compared.

Anatomy: An abundant supply of adult salmon immediately after spawning would provide material for study of internal and external anatomy and provide a model for comparative anatomy.

Physiology: The problems of an organism which has to cope with zero salinity, then adjust to approximately 3 % salinity, and finally readjust to zero again together with the problems associated with energy storage to carry the organism several hundred miles inland and uphill for three to four weeks after its last meal, should provide an insight into basic physiology.

Endocrinology: The built-in mechanism to program this living computer to swim downstream, then swim thousands of miles through the open ocean and return upstream at the right time to the same portion of the same stream from which it first started its life to correspond with the arrival of other fish, still requires much study and research. Upon arrival back at its starting place more programming is unveiled as mating behaviour finally leads, step by step, to the burying of the fertilized eggs. Then the grande finale - death.

Nutrition: The variety, quality and quantity of food eaten at various stages of the migratory cycle and the corresponding growth rates of the organism would cover most aspects of basic nutrition.

Entomology: The study of the numerous aquatic insects consumed as food, and related insects and other invertebrates that share the aquatic environments of the salmonids, would service this division of biology.

Parasitology: The many endo and ecto parasites of fresh and salt water, along with their alternate hosts, provide an approach to parasitic and saprophytic organisms.

Embryology: The eggs may be stripped from the fish, artificially fertilized and developed in incubator tanks, and studied throughout development.

Microbiology: Bacteria which may contaminate water or cause diseases in the salmonids can launch a study of bacteriology.

Ecology:

a) Population dynamics - predator/prey relationships.
b) Scavengers - as these are mainly birds we have a lead into ornithology.
c) A variety of food chains as the salmonids move through different habitats.
d) Edaphic factors of ocean, estuary and stream and the changes to accommodate each in succession.

e) Flora along the stream banks as soil retainer, food supply and cover.
f) Quality and quantity of fresh water supply.
g) Energy transformation.

Senses: Food location in muddy water. Imprinting, orientation, homing and navigation.

Genetics: Each stream has its own genetic stock which can be used to observe and test the fundamental laws of genetics. The time span is a little greater than that of fruit flies though.

Evolution: Mitosis, mutation and selection all become active verbs which shape the traits of the stock for each river, different but very similar.

Salmonids and Other Disciplines

New curricula presented today are expected to be multi-disciplinary or interdisciplinary. A study of salmonids can very certainly be broadened to include several disciplines other than biology. Some of these are:

Geography:
a) Drainage systems and ocean basins.
b) Mountain ranges and interlying valleys drained by the rivers and streams upon which the salmonids depend.
c) Land forms which regulate the amount and rate of flow.
d) Location of cities which generally give rise to people interference with salmonids.
e) Location of industries and the economy of the country.
f) Political boundaries.

Hydrology: Quantity, quality and temperature of the water supply, effects of glaciers and ground water, temperature variation and rates of stream flow all have a direct effect on salmonids.

Climatology and Weather: An important first step in the production of a water supply is rainfall. This is closely followed by snowpack, storage reservoirs and rate of evaporation.

Oceanography: Tides, ocean currents, salinity, energy exchange, changing temperatures and plankton supply all influence the cycles of salmonids.

Chemistry and Biochemistry: Dissolved gases and minerals, suspended solids, pH, nutrients, pollutants, biodegradability all influence the success and efficiency of salmonid production.

Soils: Particle size of soils and sediments and rate of silting are important to acceptable and productive spawning beds.

Geology: Composition of the substratum, especially its copper and mercury content, may directly affect the salmonids. Glacial till, degree of sorting, weathering and erosion, often determine the location of spawning areas.

Anthropology: Early native people were dependent upon the salmonids for their food supply. The salmon were sacred and revered. Salmonids are still important in the culture of these people today.

History: Early explorers followed the waterways. Salmonids were an important item of food obtained from these waterways through the Pacific North-West. Fur traders purchased the pelts of fur bearing animals which in turn were dependent upon salmonids as their food source.

Political Science: Fishing boundaries between two countries are the cause of much debate over offshore fishing rights. Agreements on fishing rights can improve harmony and co-operation between countries. Salmonids which readily transgress political boundaries both in fresh and salt water can offer an insight into studies of international law.

Salmonids and the Community

Many issues affecting the health and welfare of members of a community are directly related to the problems of the salmonids. A metaphor of the Pacific Northwest informs us that "We go as the salmon goes".

The most important of these community issues is an adequate supply of clean fresh water. Salmonids also require clean cold water. What is best for the fish is also best for the community. Therefore, through an understanding of the problems of the salmonids and by taking action to overcome these problems, and thus enhance salmonid production, community water quality must also improve. Other concerns of the community which may directly, or indirectly, affect the quality of the water supply are:

1. The disposal of garbage. Does the garbage fall directly into the stream? Is seepage from garbage disposal entering the stream?

2. The disposal of sewage. What is the effect of seepage from backyard septic tanks or cess pools? Is sewage treated? To what degree? Primary? Secondary? Tertiary? What is done with the effluent from the disposal plant?

3. Agricultural practices. What is the disposition of effluent from farm wastes? What is the routine method of disposal of dead animals? How is old farm machinery dispersed? How does plowing effect run-off? Is there excess fertilizer or insecticide carried in the run-off or through seepage? Are harvesting practices or crops themselves contributing to water pollution?

4. Mining development or production. Has construction of buildings or access roads added oil or fuel wastes to water supply? Have movements of construction materials added silt or sediment to the streams? Are the mine tailings contained to prevent washout, seepage or leaching?

5. Forestry clearing or replanting. Has cutting plugged streams or culverts? Has logging removed vegetation that anchored the stream banks? Has this caused excess run-off to change the nature or even the location of the streambed? Has removal of shade trees along the bank

altered the temperature of the stream? Have any small channels dried up as a result of this cutting? Is the slash burned? Does the slash or the burning cause the greater destruction? Does the burning remove the humus? Does it increase the danger of a flash flood after a heavy downpour?

6. Hydroelectric projects. Have man-made dams altered the character of the river? Has provision been made to assist salmonids over this dam? Were trees and stumps left to rot after the reservoir was flooded? Is the reservoir choked with snags and decaying vegetation? Have recreation areas been created? Has farm land or forest land been destroyed? Have people been displaced? To what degree has wildlife suffered?

7. Industrial projects. How much pollutant is discharged into the air? How much of this combines with moisture to form acid rain? How much falls back to earth and is washed into streams? How are radio-active materials stored? What is the method of disposal of radio-active wastes? Is the effluent from the plant dumped into the stream? Are there any toxic chemicals in this effluent? Does the effluent alter the temperature of the stream?

Interest, education, concern and action can lead to satisfactory solutions to most of these problems. A species of salmonid can be the motivator to get the community involved.

Salmonids and Education

Tangible improvements can readily be shown by counting the increase in the number of fry or smolts produced, or eventually by the increased harvest. The improvement in the quality of water can also be measured. However, the most important part of the entire program, its educational value, may not be measurable. This is not education in the old fashioned sense where the life cycle of the salmon was memorized, and the measure of the educational value was the percentage of the facts correctly regurgitated. It is education where the doing is as of much importance as the results to be obtained. It provides a myriad of opportunities for doing, for discovery, for finding out. It provides challenges for independent and original study. It encourages observation, data collecting, record keeping, sifting, sorting and analysing of data. It demands decision making and the trying out of new ideas. It requires the testing of the effectiveness of the new ideas and then the measuring, recording and analysing start all over again.

The first series of problems is to define baselines for each of the many parameters of the streams. What is normal for each stream? How does one stream differ from another in each parameter? How much does a stream differ from the conditions considered as optional for salmonids? What variation is tolerable in these conditions? How different are the stream conditions from those considered tolerable? How do stream conditions change from year to year? How do optional stream conditions vary for different species of salmonids? A variety of instruments, devices and much ingenuity are used to obtain the necessary measurement of stream parameters.

There are stream investigations that can be done by 10 year olds or grade four children. These investigations can readily increase in complexity as the children return in successive years until through secondary school.

There is much detective work that can be done in locating causes for temperature changes, rate of flow changes, changes in turbidity or changes in chemical composition. Many mini-engineering problems must be solved as the students work to collect their data.

After the mastery of techniques for measuring the many parameters of the stream then many comparisons are possible following simple manipulation of streambeds. For example, after moving a log or placing a boulder, which parameters have changed and by how much? What effect does a small dam have on stream flow and on water temperature?

Parameters measured by the students would include stream cross sections, rate of flow, pH, temperature, salinity, dissolved oxygen, total dissolved solids, organic material, coliform bacterial count, chemical composition, turbidity, rate of sedimentation and size of gravel.

An Action Program

School classes, groups of citizens of clubs should be organized to work on a single stream. If preliminary work is done, fisheries officers and community education officers of S.E.P. will offer advice and technical assistance. Permission of landowners is mandatory, and a letter of intent to Ministry of Fisheries is very desirable before action commences. Then:

Adopt a stream.
Walk-through of area.
Construct a walking trail.
Inventory flora along bank.
Map main stream and tributaries.
Map community, homes and buildings.
Map farmland adjacent to stream.
Map nearby industries.
Locate and plot sewage units, septic tanks and outfalls.
Inventory aquatic invertebrates.
Map bank erosion.
Map gravel beds.
Start daily records of stream parameters mentioned above.
Remove debris and obvious ostacles to stream flow.
Create pools, shelter and shade.
Adjust and manipulate flow to approach optimal conditions for salmonids.

Place fish box and run bio-assay for 96 hours.
Introduce or rehabilitate fish.
Build and install incubation boxes.
Construct rearing channels.
Construct spawning beds and channels.
Add nutrients to lakes.
Construct fishways and ladders.

Salmonid Enhancement Program (S.E.P.)

In 1977, the federal and provincial governments in cooperation set up a multimillion dollar, multiobjective, 20 year program to study all facets of salmonid enhancement. This was to be translated into action and where feasible implemented. School groups, clubs, community groups and individual citizens were invited to participate in the project. A program of formal and informal education was to be included.

Financial support was to come from public funds, but in the form of an investment, because it is estimated that the money will be repaid many times over as the annual harvest increases. A levy will be paid by sports fishermen who will benefit from the improved recreational facilities and increased yield.

Researchers, technologists, fishermen, educators and communities are all involved in solving the problems associated with improving the harvest and its processing.

Salmonids and the Teacher

Prior to the introduction of the S.E.P. some teachers in the province took full advantage of the educational opportunities available through activities based on salmon. One of note was that carried out by Allan Wootten in Terrace, B.C. With the blessing and cooperation of the local fisheries officer, eggs were stripped from mature Coho salmon. After fertilization these eggs were placed in several baskets which were then buried in the gravel streambed of a spawning area. Some eggs were taken back to the classroom and incubated in an aquarium under controlled temperatures. Approximately each week throughout the winter the class returned to the spawning stream to retrieve one of the buried egg baskets. These eggs were brought to the classroom to compare their development with the monitored eggs. Numerous mini-engineering problems had to be solved to adapt rearing technology to this educational activity, but this in turn increased the value of the exercise to these junior secondary students.

Other teachers have made extensive use of stream studies centered around spawning runs. Since the implementation of the S.E.P. much help is available to teachers, classes of students and other organizations or individuals who require assistance. Two specific programs have recently been published which provide factual information and practical ideas for involvement in the S.E.P.

F.A. Gornall

1. Salmon Streams and Salmonids - A Handbook for Water Quality Studies. Dyckman, et al. Unesco Transactions Project.

> "This is a manual for those who have the interest and desire to undertake a stream study so that we may better understand the efforts of local technicians, scientists and experts to restore our water quality, our streams, our habitat and our fisheries. It is for lay persons, educators and technicians alike."

The manual provides a program of first-hand activities involving children with streams. Four divisions each deal with an aspect of major importance in enhancement. These are:

a) Small stream study with community and school.
b) Biology of salmon.
c) Small stream enhancement projects.
d) Salmon culture.

The manual contains all the routines to test for baseline parameters of a stream.

2. Salmonids in the classroom. A program developed and produced by Dr. Glen Sinclair and Associates for the S.E.P. This seven volume set of reference materials, audio visual aids and student activities is available through the British Columbia Teachers' Federation Lesson Aids Division. It brings together all the facts known about salmonids and describes dozens of classroom tested activities related to salmonids. Teachers are being introduced to these materials through half or one day workshops.

Salmonids and the Future

The S.E.P. program will continue for many years. A.B.C. Task Group, a broadly based advisory group of thirty members, is a major source of public input. Its members represent the varied interests of the community.

"S.E.P. Update 78" is a series of public meetings held in 20 communities throughout B.C. The concerns and suggestions voiced at these meetings pass through the Task Group to the S.E.P. Board for action.

Periodically a Salmonid Newsletter, free on request, informs of new projects, techniques or teaching ideas.

S.E.P. mails an annual report summary free to every school in the province.

Summary

Salmon production can be improved on most streams whether they are in pristine condition or affected by man. Spawning can be improved by controlling bank erosion, adding gravel, removing small obstructions and clearing debris and garbage from the stream. Ensuring that enough cover, instream backwaters and eddies remain can improve the juvenile production of salmonids. This kind of enhancement requires little equipment, some labour and a practical commonsense knowledge of what salmonids need.

We as educators should encourage teachers to use this natural phenomena as an educational tool. We should help them, their students and other interested groups in working closer to their environment and moreover to move closer to an understanding of their environment.

BIOLOGY COURSES FOR NON-SCIENTISTS AT THE TERTIARY LEVEL

by R.A. Kille

Our Commission is charged with producing practical programmes in aspects of community education. This paper will be confined to our workshop on biology courses for non-scientists at the tertiary level. I propose to write a discussion paper, raising the constraints and possible objectives of such courses which I hope will make some contribution towards our task. The primary constraint on our task is that these courses are urgently required around the world. We must produce some practical programme based on whatever experience is currently available, leaving further experimentation and refinement to be developed within operational courses. I believe enough thought has already been given to course content, organization and philosophy to enable useful biology courses for non-scientists to be launched in any part of the world. An aspect of these programmes which needs most attention in all regions of the world is the expertise to train teachers in the new approaches which these courses will need if they are to avoid becoming diluted traditional introductory courses in biology. A new pedagogical approach takes a long time to significantly influence the mass of any teaching population, but nowhere more-so that at the tertiary level. Tertiary level staff are selected for their originality which is closely linked to a widely independent attitude, including change in pedagogy which they regard as secondary to their research function.

The Need for Biology Courses for Non-Biologists at the Tertiary Level

I should justify the statement above that biology courses for non-scientists at the tertiary level are urgently required, not only in an attempt to satisfy cost-effective criteria, but because a consideration of the need for these courses may give us some guidance for objectives, content and structure of the courses. In 1956, the author and scientist C.P. Snow introduced the problem of the 'two cultures' and developed his concern further in the Rede Lecture in 1959. He saw the intellectual life of western society increasingly being split into two polar groups. -- 'Comparable in intelligence, identical in race, not grossly different in social origin, earning about the same incomes, who have almost ceased to communicate at all, and who in intellectual, moral and psychological climate had - little in common.' 'On the one hand the non-scientists have a rooted impression that scientists are shallowly optimistic, unaware of man's condition. On the other hand, the scientists believe that the literary intellectuals are totally lacking in foresight, peculiarly unconcerned with their brother men, and in a deep sense anti-intellectual, anxious to restrict both art and thought to the existential movement.' Many of us will have direct experience of this lack of understanding between scientists and non-scientists so vigorously expressed by Snow. The incomprehension of Science gives, much more pervasively than is generally realised, an unscientific flavour to our total culture, and that unscientific flavour is all too commonly becoming anti-scientific. The traditional culture, Snow argues, is the one which still manages the western world and wishes the future would go away. It is the non-scientific graduate that still dominates decision making in housing, education, health and conservation, both in central government and locally. There is urgent need for communication and co-operation between the two poles of culture to improve the human condition.

Tradition in all things is remarkably persistent, and tradition in major educational attitudes is no exception. Intense specialization at the tertiary level has dominated the secondary school programmes so that it is common for the science or non-science decision being forced upon pupils of 12-13 years of age. Clearly the secondary level is where the dichotomy between science and non-science should be stifled, forcing the tertiary institutes to change their teaching to cope with a broadly cultured intake. Until this occurs there is an urgent need for science courses for non-scientists, (and equally important, non-science, traditional culture courses for scientists) at the tertiary level, from which the majority of senior executives are recruited.

There is no need to argue the case for biology as the most suitable course for non-scientists at the tertiary level because that is our remit and our interest. However it would not be difficult to defend a case for biology. Biology is the most inclusive science, and biological concepts are at the root of much of the present human predicament. Food shortage, health, housing, pollution and social organization all include a large biological component.

Separate or Common Courses for Tertiary Level Biologists?

The assertion that non-scientists should experience scientific rigour raises the question of teaching the non-scientist alongside the scientists. A panel of biologists in America recommended that biology majors and non-majors should be taught in a common course (Commission for Undergraduate Education in the Biological Sciences Report No. 19. Biology for the Non-Major 1967). This surprising recommendation was supported for three reasons.

1. Many small colleges do not have the staff or facilities to offer two different introductory courses..
2. If a student in the non-majors course should become interested and wish to become a biology major he would have to take a further introductory course for majors.

3. When a separate course for non-majors exists side-by-side with a course for majors, the former often becomes a diluted version of the latter.

Undoubtedly one or more of these situations will have significance for very many institutions. In many tertiary institutions, and not only those in developing countries, there would be a lack of trained staff and other teaching facilities to physically provide two courses. There would also be the problem of finding enough staff with the aptitude and motivation to devise and maintain a separate course for non-scientists. However biologists can no longer shirk their responsibility. The future of biological science itself and the contribution it must make if health, happiness and the full spectrum of human culture is to continue depends upon all of mankind understanding and being wholly sympathetic towards science and especially biological science. We cannot leave it to non-scientists to glean what they can from courses for scientists. There must be a positive and direct action by biologists in the form of courses specially designed for the most effective education of non-scientists.

Moreover I believe that where the motivation and expertise exists, courses sensitively planned for non-scientists would soon be seen to be the kind of course that should be provided as an introductory course for majors in biological science. Many present introductory courses are lumbered with the accretion of scraps or biological knowledge of the last hundred years. Such introductory courses obscure the methods of science, which should surely be the primary aim of science courses. Trained in the techniques and philosophy of science, the details, complexities and relevance of the higher levels of tertiary courses would be more readily and profitably assimilated. Meanwhile the need to educate non-scientists in the ways of science cannot await the re-education of conservative tertiary-level staff. We must press on with separate courses for non-scientists.

Course objectives

Bearing in mind the constraints discussed above, what are the achievable objectives? It would seem from the discussion in the working party on this topic in the Chelsea meeting that there are three general objectives which would find common agreement. These are:

1. Science as a culture -- the intellectual and conceptual substance of science.
2. The methodology of science, its validity and limitations.
3. The role of biological science in finding solutions to contemporary health and environmental problems.

Ad 1. If we accept that it is our responsibility as biologists to help bridge the gap between the two cultures, then one of our objectives should be to present science as a culture to non-scientists. It would not be difficult to show these students that science is a culture akin to their own specialization. Just as music, literature and painting (all fundamentally biological phenomena!) explore and communicate the nature and significance of ourselves and our environment to enrich our lives, so can science. Examples could be chosen to show the influence science and the scientific process has had on freeing the human mind from religious and moral conventions, with all too frequently their accompanying social injustices. In presenting this aspect of the science it is essential to show that scientists do have a sense of morality.

Biologists are fortunate in the task of presenting science as a culture because it is an easier task than it is for physical scientists. The biological concepts of evolution, heredity and variation have freed mankind from all kinds of dogmas and conventions about himself and other living things, and profoundly affected his social and industrial organizations. The intellectual and conceptual aspects of science will be of interest to all non-scientists if presented sensitively, whatever their reasons for taking the course, and whatever their future concerns may be.

Ad 2. The second objective should remove much of the mystique which has been built up around science and the scientist. The 'science can do anything required by mankind' attitude, which fails when the problem is beyond the present capabilities of science leads to distrust and disillusionment with science and a move towards mysticism and irrationality. The attitude that science itself has some kind of magic process, peculiar to science should be dispelled. It should not be difficult to show that the great strength of science is that it is available to all who perform the disciplined operations by which it functions. Maybe great minds are required to produce seminal ideas, but science is mostly carried forward by a large number of lesser intellects evaluating and adding to the total body of science.

The course must provide the opportunity for students to carry out short investigations for themselves and study the methodology of selected pieces of experimental research. Here again biological science has an advantage over the physical sciences. The phenomena of variation and adaptation give another dimension to biological scientific inquiry. Skill acquired in analysing the effects of these phenomena can be readily transferred to human social and economic fields.

In pursuing this objective the course must emphasize the limitations of scientific enquiry as well as its capabilities. Furthermore, it is important to avoid on the one hand giving the impression that the scientific method is a rigid process, and on the other that it is so undefinable that it cannot be usefully discussed and demonstrated.

Ad 3. The third objective is likely to be demanded by students taking the course if it is not already planned. It is here that the biology of man and other living things, their relationship to each other and the limits of the Earth's resources can be developed in the context of some of the pressing problems of our day and the immediate future. While the emphasis in the course should be on human biology, non-scientists should not be given the impression that this is the whole of biology. It will be necessary to underline man's relatively recent appearance upon the biological scene, and the serious effects his activities are having exponentially on the environmental equilibrium established long before man appeared on the scene.

The summary of the objectives of the 'Science in Society Project', sponsored by the Association for Science Education, given by Dowdeswell, et al., seem to me particularly appropriate:

- to promote understanding of the nature and limitations of science,
- to encourage a realization that the use of scientific knowledge can be both beneficial and detrimental to society and the environment,
- to foster the idea that the Earth's resources are finite,
- to develop an understanding of the importance of reasoned decision making, taking into account all relevant constraints; also a recognition that moral considerations must be borne in mind.

Structure of the Course

I consider the aspect of course planning to be the most critical and the most difficult. Provided it is accepted that a separate course is necessary for non-scientists, the objectives of the course need not be very contentious and the content chosen to achieve these objectives can within limits, albeit fairly generous ones, vary from one region and institution to another. The framework of the course is, however, very important if we are to achieve the main thrust of our objectives which is the order and cohesion within biological science in spite of the great complexity of the ecosystems and the variety of levels of organization from the molecular to the population.

How do we present the impact of biological science on the life and thought of mankind? How do we show that biological science is an expression of the human creativity and that scientists show all the human foibles in their acts of creation ('The Double Helix-Watson') throughout history? How do we give biology itself the full treatment from the molecular to the ecological? And how do we include discussion of solutions to the problems of starvation, pollution, over-population and human aggression?

Forty years ago introductory courses were organized around type specimens which displayed different levels of organization. To this framework was added separate topics such as genetics, embryology, alternation of generations and evolutionary theory. Work in molecular biology, biochemistry and physiology served to emphasize the unity underlying the superficial diversity of living things from microbes to man, so that 10-20 years ago there arose a wide spectrum of course structures, based on sub-divisions of biology viz. physiology, ecology, evolution, molecular biology.

Other strategies centred round the scientific inquiry approach, (BSCS, Nuffield).

We should give some consideration to all these strategies in our discussions. Most of them can lead to sound courses when skilfully used, but all of them tend to lead in time to courses in which the biology becomes compartmentalized. There is little, if any, integration of the units and the student fails to see the whole wood for the trees. In an introductory course for biological scientists at the tertiary level this may not be disastrous because their immersion in biology for a further two or three years may give them a view of the wood. For non-scientists we need a structure which will firmly link the components of biological science, but at the same time not overwhelm them with the complex inter-relationships of organ systems or organisms.

Some may advocate a strong bias towards human biology as the integrating theme, but there is a danger that this approach may not give a fair and objective view of biology itself. The strategy for a human-centred structural approach of Dr. Kattmann of IPN is very attractive but it does not seem to leave much room for other aspects of biology in a one year course. The cybernetic approach of our host, Dr. Schaefer, in the same paper has the advantage that it provides a system which could be applied to the whole field of biology. Its particular merit is that it can also assist the teaching function.

It is particularly fortunate that we are meeting in IPN and therefore there will be the opportunity to get our hosts to develop their systems for our own education.

Methods

The only substantial point that needs to be considered here is whether these students should be exposed to laboratory work. In my opinion this is essential if the art of scientific inquiry is to be fully apprehended. Carefully planned investigative exercises will enable students to find out for themselves what one does to make valid observations, how to derive generalizations and from them generate ideas which can be tested. As Hanson, 1970 says, the student -- 'must learn what has been done so that a person is willing to say and assume public responsiblity for what he has said.'

Content and Evaluation

If we set ourselves clear objectives and find a structure for the course which will achieve our objectives, the content of much of the course will become self-evident. Moreover, some degree of latitude in the content must be allowed to enable the course to have some relevance to the local interests of the students and presenting staff. It may be helpful however, for us to prepare a list of suitable reading material and other teaching aids.

As far as evaluation is concerned, if one emphasis in the course is on the nature of scientific inquiry, then evaluation based on rote learning and recall is out of place.

References

Commission for Undergraduate Education in Biological Sciences. (Washington D.C.) Report No. 19. 1967. 'Biology for the Non-Major'.

DOWDESWELL, W.H., 1979. European Journal of Science Education. Vol. 1 (1) pp. 51-56. 'Science and Technology in the Classroom'.

HANSON, E.D., 1970. CUEBS News 1-6. 'Biology for the Non-Biologists; aims and issues'. Commission for Undergraduate Education in Biological Sciences, Washington, D.C.

KATTMANN, U. and SCHAEFER, G., 1976. J. Biol.Education. Vol. 10 (3) 139-147. 'New approaches to restructuring school biology'.

LEWIS, J., 1977. Science in Society Project (Association for Science Education, U.K.).

SNOW, C.P., 1959. The Rede Lecture. 'The two cultures and scientific revolution'. Cambridge University Press.

WATSON, J. 'The Double Helix'. Weidenfeld and Nicolson. 1968.

SOME THOUGHTS ON EDUCATING BIOLOGY TEACHERS FOR COMMUNITY DEVELOPMENT

by C.A. Quiroz

Introduction

Although community development is a sort of task for every member of the concerned community, in terms of education it seems quite natural to think of science teachers as the best qualified individuals for some kind of action along these lines within curriculum implementation.

Biology teachers' action, in particular, seems to be very relevant in this field, because while doing formal teaching, they have to deal with topics related to the understanding of life, in all of its forms including human life. Topics such as interactions, physiology, ecology, energy flow, and so on can be used to study real phenomena and problems occurring and affecting the community where the teacher's influence is being exerted. However, I would like to stress that community development programmes, implemented throughout the school activity, call for interdisciplinary commitment,and therefore all teachers in any school community as well as other staff members and students, are to be aware of their responsibility as social community development promoters and effectors.

As Professor Kelly suggests in his working document, curriculum planning for biology teaching (and for other subject matters as well) ought to be based on community development problems and issues. In other words, the class motivation is to be found in real life examples, and therefore the actual teaching of biology is to be conducted primarily aiming to educate children to become active members in the development of their own community and also be prepared to preserve and if possible improve its values and resources.

C.A. Quiroz

I feel myself very much inclined to entirely agree with Professor Kelly's working document, which I think provides us with a fairly good coverage of the many aspects to be taken into account in dealing with education related to community development. Out of these aspects I would like to address myself to the biology (and others) teacher role in this enterprise; and, therefore to concentrate attention on preparing him (her) to become a 'community development promoter'.

The Biology Teacher seen as a Community Development Promoter

I can hardly concentrate my attention on the education of the biology teacher alone. Although I myself am, by formation, a biology teacher, over the past few years I have become more and more interested in integrated and/or interdisciplinary approaches to education. Therefore, I think a special point needs to be made regarding the education and preparation of those people who by reason of their commitment are to be influencial on the new generations and in doing so may become the energetizers and promoters of community development. I mean teachers, of course; and in our particular case biology teachers.

In considering this point, I am thinking not only in devising for them the right kind of scientific or science-related subject matter in curricula, which can be relevant to the various study issues as suggested by Professor Kelly, and many similar others as well; but also to include those matters related to human interactions as communication skills, interpersonal and social concern, self-development, and leadership attitudes, amongst others.

What I mean here is that the teacher, no matter the subject contents he (she) will be dispensing in the classroom, in addition to acquiring a good scientific background and the understanding of what science is about as well as its potential value as an instrument for community development, must also learn to develop a particular personality to become an activator of both the school community members (staff and pupils) and the social community in which the school is inserted and its members are also a part of. And this can not be achieved just through formal disciplines.

I like to think of teachers as playing a dual role, being at the same time teachers and parents. These two roles are very much related to eac other as both of them are in turn related to the welfare of children. As human beings, teachers either are or can become parents; it is the nucleus of the basic community cell: the family. Even if for any reason or the other, they are not parents, they still could have a special feeling or preference for a little nephew, cousin or brother, who is most belovec to them. In general, I think, every one of us as a parent wishes the best for his (her) children. And the best is not just a comfortable home, love, and other personal facilities and commodities, but also a good education and the best possible social environment as well.

So bearing this in mind, and by becoming aware of this dual personality, being a teacher means much more than just being a transferrer of knowledge and therefore acquires a greater dimension. The teacher is not any more a simple knowledge conveyer; he is to become a leader, a promoter, an energetizer, an awakener of the skills, abilities, and willingness of both the children and other members in the social community, to encourage and become truly involved in the development of the community towards the acquisition of those qualities which could make life really enjoyable for everybody.

Hence, I would suggest that in addition to Professor Kelly's five suggested projects for action we ought to engage ourselves in developing ideas for a programme to prepare teachers, and specially biology teachers, to foster and encourage the development of those suggested projects, and consequently to become real community development leaders.

Now, focusing this idea within the framework of this workshop I may suggest that biology teacher education could be taken as a sort of model for trying such an approach. Actually I hope to gain some inputs which could help in the implementation of some action along these lines both for pre- and in-service teacher education at home.

Possible Ways of Action

From a more general stand-point and trying to relate these ideas to the contents of the working document, I would like to refer myself to some efforts made at home and elsewhere along these lines or which can in any form be related to the main objectives of the workshop.

One possible way of action are seminars and/or workshops (like this one) in which representatives from organized bodies of the local, national and international community could contribute by giving up-dated information and ideas on the various problems affecting the community concerned, and on the available resources, as well as what is expected from the various statements of the community towards the improvement of the quality of life.

These meetings could consider and propose for action, feasible projects to be developed as school-based activities. An example, in addition to the one we are attending to, could be the one held a few weeks ago in Lima, where a number of science and mathematics teachers, coming from various secondary schools in the country, with the help and contribution of UNICEF, Ministry of Education, Ministry of Agriculture, National Bureau of Evaluation of Natural Resources, National Institute of Occupational Health and the local Information Centre of the United Nations, gathered to discuss and consider the role of the teacher in environmental education (and community development).

Another example is provided by the 'Peru Pilot Project on Environmental Education in Secondary Schools', which steemed off from a Regional Meeting on Environmental Education, convened by Unesco in Lima, Peru in 1976.

Another way of action is that of devising and carrying out interdisciplinary projects for the study of a particular problem in the environment, such as the 'KARE Project' developed at Montgomery County in Pennsylvania, USA. This may involve both secondary schools, teacher training centres and local universities, as well as other local or/and State organizations.

Before the members of the community may become interested and involved in developmental projects, it would be necessary to establish a sort of link between them. In other words it would appear necessary to open channels or means of communication. Of course these could be the mass media. However, they tend to be too impersonal and therefore something which could bring them into closer contact and let them to meet each other and share each other's concerns and queries, is to be developed. To my mind, these organizations already exist and we only need to use them as far as they can be of any use. These are the Lyons, Rotary and similar clubs. They can be interested to become involved and they do become involved, in educational projects.

In addition to these a 'Community House' could be established where members of the community can freely meet without the need to being affiliated to any kind of society, and discuss local problems and plans for feasible solutions. An example of this idea, comes again from the Community Education Nucleus No. 02-04 in Lima, Peru, where one of such a house is being implemented now with the assistance of Unesco and the local authorities.

SOME BARRIERS TO EFFECTIVE BIOLOGICAL EDUCATION

by F.A. Gornall

Some barriers to an effective understanding of biological relationships are constructed early in the life of the child and unfortunately often remain throughout the life span of that individual. Other barriers are erected as the child moves through our school system. A few barriers are surmounted during adolescence but others arise to take their place. Reinforcement is constantly being applied. Thus, many barriers are carried into adulthood and provide a mental block to logical reasoning about rational thinking on biological problems.

One of the first barriers is built into the child at an early age with his early exposure to nursery rhymes. An attempt to criticize nursery rhymes evokes the response, "What's wrong with nursery rhymes? Children love them!"

Children may love them but children are much more likely to love the attention they receive as the result of an encounter involving the nursery rhyme. The warmth, the touch, the relaxation, the smile, the complete attention given by the teacher of the nursery rhyme are what truly affect the child. The warm response and indication of approval at the child's quick mastery of the simple rhyme reinforces the learning situation. The child may not understand the words but repetition earns his reward and the rhyme becomes entrenched. All that is probably good, an essential part of the child's development, but the nursery rhymes per se, what do they do for a child?

How many children have a lifelong fear of spiders because of the "Little Miss Muffet" story? Spiders are a neglected area of study. Realistic spider stories seldom appear in childrens' readers. Spiders are ignored in most early textbooks. To what degree can Miss Muffet's fear be held responsible?

What image does a child receive from another nursery rhyme:

Mary had a little lamb,
its fleece was white as snow?

It is almost a traumatic experience for the child a few years later to run his hands through the thick, kinky, dirty, oily, grey, warm, matted, smelly "fleece" of a sheep. But what of the child who is deprived of the opportunity to run his fingers through the wool on the sheep's back?

The problem arises not directly from the nursery rhymes themselves but rather from the lack of direct experiences which counteract images inspired by the nursery rhymes.

A second barrier is that of stories for young children which impress vivid, but often terrifying, steriotyped images of certain animals. For example, "Little Red Riding Hood". Wolves remain long toothed villains in many other stories and in real life because Little Red Riding Hood's wolf had developed a taste for grandmothers and little girls. A second example is that of "Goldilocks and the Three Bears". How many adults are terrified because they might come face to face with a bear if they were to take a walk in the woods? It would be unfair to say with certainty that Goldilocks was responsible for people's attitudes about bears but adults seem to be about equally divided into two camps as far as attitudes towards bears are concerned. In most cases, very little bear information has been given to these adults since childhood. One of the two groups ignores all warnings about feeding or leaving food available for wild bears. They treat bears as they would the family puppy. The second group believes that wild bears will seek them out to devour them. Both extremes are unreal and yet these attitudes persist in spite of literature available on behaviour of bears. Should that cute and cuddly Teddy Bear from the nursery share the blame?

The early impressions which children receive about animals allow them to categorize animals into three groups, (a) "bad" animals, (b) "good" animals, and (c) animals that don't matter. Once again, these impressions remain into adulthood. Factual information and direct contact with animals is lacking. The school programme does not include enough stories to counterbalance such early favourites as "Alice in Wonderland" and "Burgess Bedtime Stories". We observe unwarranted persecution of the "bad" animals such as hawks, owls, and eagles or coyotes, foxes and wolves. Members of the "good" group are deer, rabbits and squirrels or ducks, geese and song birds. All are colourful, melodious, or good to eat. The third group can be relegated to non-importance because they are "only" crows, magpies and their ilk, or rats, mice or gophers. Of course, everybody knows that these are all a nuisance and of no use to anybody.

Anthropomorphism plays a large part in childrens' aquaintance with animals. In its earliest stages, it manifests itself through animals wearing clothes and talking to one another, or by talking and listening to humans. Sometimes these animals live and act like humans. Most of the time this is only a picture or storybook world, but even into adult life it is still accepted as cute and funny. This can be ascertained simply by glancing at the "comic" section of our daily newspapers. Adults realize much of this is childish but continue to promote it. Disney's "Donald Duck" and "Mickey Mouse" are well known characters in this group.

A much more serious aspect of anthropomorphism is that perpetuated even by experienced biologists in a more subtle way when they refer to animals and even plants as liking a certain environment, or preferring a certain situation. Even though sometimes the biologist may know the difference, he or she reinforces the earlier impressions accepted and seldom contradicted.

Teleology is another barrier to biological education. Such expressions as "Beavers have sharp teeth in order to cut down trees" are common in our daily conversation, and appear very frequently in our literature. In a quick survey of several science text-books it only took a few minutes to locate several such teleologic statements in each book. These books ranged from early primary through junior secondary science courses. Even university texts are not free from the use of teleology.

Teleology enters the child's world at an early age but becomes most deeply entrenched in adults. Adults often use teleology but many are unaware that they do so. Exposed to such statements as "birds fly south in order to survive the winter", or "giraffes have long necks so that they can feed upon higher foliage", or "leaves change colour in autumn to brighten up an otherwise drab landscape", children have teleological reasoning reinforced and perpetuated.

If one is aware of the misrepresentation or misunderstanding created by the use of teleology and has the audacity to comment upon it then that person is immediately accused of indulging in semantics. Arguments over semantics are semantic in themselves and do no solve any problems. Meanwhile the teleology remains as a barrier to reasoning, a block to thinking.

The perpetuation of a myth is yet another barrier. Myths about animals are always popular with teachers and are therefore perpetuated in spite of the evidence which belies them. Myths tend to romanticize animals, ie., the unicorn, fiery dragons, and suicidal lemmings. There is also an implication of mystery and the unknown which appeals to many adults. Perhaps this is what helps to keep the story alive. Many fantasy cine films have capitalized on this in the entertainment field and unfortunately have been the sole source of information or misinformation about the animal.

How deeply entrenched a myth may become is perhaps exemplified by an anecdote about arctic lemmings. A group of non-science university students was asked to write a paragraph or two about lemmings of the Canadian arctic. In response 28 of 31 gave a fair account of the mythical but non-existant suicidal marches to the sea. Early stories of suicidal lemming migrations in Scandinavian countries reinforced by the Disney film "White Wilderness" perpetuated and transferred this myth to Canada.

Mother Nature must also share the blame for shoddy thinking about biological problems. This personification of a pseudo god-like being who controls all living organisms is a barrier to rational thinking, ie., "What is the point of me thinking about it when Mother Nature looks after own?" When a phenomenon is not understood then the problem is rationalised by blaming it on to Mother Nature. It effects closure.

"Mother Nature" is used more in communicating with children. For adults the "Mother" is dropped and "Nature" continues as the organizer and controller, as the planner and discriminator.

Reference to "Nature" leads eventually to "Balance of Nature". This is still another barrier that prevents students from seeing the world as it really is. A frequently heard statement is, "We must protect the delicate balance of Nature". What really are we trying to protect? The statement implies that we should prevent change or practice preservation of what we have. A challenge that I have often given to secondary biology students is to mark off an area of a couple of square metres and to PREVENT change. No one has yet suceeded. Do we need the term? I believe it is an anachronism from the dark ages and with our increased awareness and understanding of ecological relationships the term is no longer necessary or even useful and we could make better progress if we stopped trying to wish away problems by false reliance on a cliche'.

Adaptation to Environment is yet another barrier to slow down biological reasoning. Frequently it is tied together with teleology. There is little argument that adaptation does in fact take place but gross misunderstanding occurs because time, much time, is required before these adaptations become established. Many students expect adaptations to take place almost instantaneously before their eyes. Again questions asked of students indicate that this is so. An example: Question: "If I place a brown snowshoe hare on a patch of snow what change will take place?" Many changes may take place but 19 out of 20 answers are: "It will turn white" or "It will adapt to its white environment by turning white". A couple of simple rules might help to overcome confusion. These are: Short time changes of an individual animal or plant should be referred to as adjustment. The term "adaptation" should be reserved for long term changes in species, and not used for individuals, and the third, the organism that fits the environment best is the one which is left after all the others have been culled out, ie., the negative results of natural selection. Finally Mitoses over several hundred generations is the basis for adaptation.

Sentimentalism is our next barrier. The Easter Bunny is probably our worst enemy in this category. Frequent reference to candy rabbit eggs associated with Easter must also create biological confusion. Bambi and Rudolph the Red-nosed Reindeer are starts in the animal personality parade, followed by Santa Claus' reindeer. Again there is seldom a word or factual experience to follow.

Special events which could be real motivators for animal study are wasted in non-sequitar activities of a trivial nature. Included here would be the pumkins, black cats, spiders and bats of Halloween, all drawn, painted and cut out as symbols of "All Hallow's Eve" but although the exercises are repeated annually and at several age levels, seldom if ever, does any biology become apparent.

A recent survey of 30 primary school classrooms to determine the extent of biological displays indicated:

Only three rooms with live caged animals.
Three rooms with displays of pictures of animals.

Four additional rooms with a casual picture of an animal such as a calendar picture.
Fourteen rooms with living plants of which twelve were decorative and two were active children's projects.
Twenty-one rooms with animal caricatures or personific ation. These had reference to sports teams such as cougars, lions or panthers, were the names of class groups such as toads, caterpillars and locusts, were stylized animals as backdrops such as chicks, fish and frogs, or the caricature included a mouth bubble carrying some special, printed message.

The barrier here is the lack of effective material to counter-act the ineffective, erroneous and misleading one.

CONCEPT FORMATION IN SCHOOL BETWEEN SCIENTIFIC AND COMMUNITY DEMANDS [1)]

by G. Schaefer

1. Introduction

A great deal of emphasis has been laid in science upon the development of a coherent logic system of scientific concepts forming an epistemic structure within scientists and learners, which suits to empirical reality on one hand and to some theoretical reality (philosophy of life, "Weltanschauung") on the other. While scientific demands, however, focus upon *invariances* (regularities, intersection sets) in an objectivated world and try to spread the use and designation of those invariances as "normed" concepts all over the scientific community of our globe, many other communities (national, local, professional, religious ones) as well as single individuals, tend to go into the opposite direction and develop a marked *variance* of meanings for the same words. We may contrast both tendencies by the following arrangement:

scientific community	other communities; individuals
- tries to eliminate the subject component from experience	- emphasize the subject component in experience
- concentrates interests upon the "objective" world	- concentrate interests upon the "subjective" world of groups and individuals
- seeks a high invariance of meanings in concept formation on an international level by definition	- allow a certain variance in concept formation within the community; furthermore, seek a high variance between different communities in order to accentuate the differences.

1) Extract from a paper presented at the International Seminar "Cognitive Development Research", Sep. 1979 at the University of Leeds, England.

- restricts the meaning of a concept to a small number of logic relationships ("logic core", see Schaefer 1979).
- enrich the meaning of a concept with a highly differentiated network of personal associations ("associative framework", see Schaefer 1979).

In science teaching at school we have to ask ourselves, for which kind of community we are going to prepare our students. In previous decades many teachers were convinced they had to train students not only in science (in scientific thinking), but for science; students were regarded as future scientists, although everybody knew that only a very small portion of the students later on really became scientists or science teachers. Today more and more teachers and educational planners change their mind with respect to the purpose of science teaching; they favour the approach "science teaching for the non-scientists" (see Kallen, 1978) emphasizing the significance of selected subject matter and scientific processes for education of everybody, i.e. of the future politician, lawyer, mechanic, artist, nurse, housewife.

The following paper is based upon this approach. One implication will be that concept formation in science teaching has to find a compromise between the two opposite demands of communities listed up above. We have to develop teaching methods that allow training concepts both in their normed semantic function (the interpersonal scientific meaning, which is highly context-independent), and in their pragmatic function of personal appeal in the ever-changing context of everyday life (the person-resp. community-dependent meaning). In other words: We have to train both the "denotative" and the "connotative" side of concepts (Carroll 1964).

2. A Compromise: Inclusive Thinking

2.1. Scientific thinking in modern biology. Biology comprehends itself as the "Science of Life". It pursues, like all natural sciences, the aim to describe its objects - here: the bio-systems - as objectively as possible, to classify them, explain them causally, and to make them measurable, calculable and their behaviour predictable by quantification of variables.

On this way part of biology changed in the last three decades into a chemical modification called "molecular biology", which allows to reduce some of the well-known phenomena of life (e.g. heredity, mutation, growth) to molecular processes and to describe them on this level in the precise categories of chemistry.

At the same time another part of biology passed through a countermovement: it did no longer concentrate upon the substantial (biochemical) aspect of biosystems, but focussed to a growing degree on aspects of microstructure and organization. Research on biomembranes, on endoplasmic reticulum on the morphologic side, and on cycles, particularly regulation and escalation cycles (Schaefer 1972), on the functional side increasingly allowed insights into the finest *static* as well as *dynamic* structures of biosystems. Parallel to a molecular biology there was a development, for example, of a "cybernetic biology", which brings in more precision into biological concepts from the mathematical formalization side leading to further quantification of biological variables and thus to a better assessment of the behaviour of biosystems (see the fundamental work by Ashby 1956).

And a third movement can be indicated: the ecological movement. This movement, however, did not lead to a more "precise" description of life phenomena in the physical, chemical or mathematical sense, but to a clearer understanding of the genuine *biological* character of biosystems. It was brought to mind more consciously by this movement, how much life processes are dependent from "outside" factors and are interwoven with them, or to put it briefly: how much they are *characteristics of larger, more comprehensive systems* ("system properties") and not so much properties of single elements like organisms (cf. v. Wahlert, 1977, Kattmann, 1977). Scientific "precision" - in the course of this development - was no longer understood as the limitation to a strict (and narrow) category of description, i.e. to an *exclusive thinking*, but as an enlargement of scientific objects to more comprehensive ones (ecosystems, biosphere) and as an increase in the number of aspects of consideration (geographical, metrological, social, political, economical aspects). This is illustrated by the humourous German ecological slogan AHMAZ, which sounds Turkish but means: alles hängt mit allem zusammen (everything is interdependent with everything). This kind of thinking has been called *inclusive thinking* by the author (Schaefer 1975). A comparison of the exclusive and the inclusive thinking shows which changes science teaching has to undergo from a "reductionistic" to an "integral" form of teaching in order to become appropriate to the specifity of life (cf. also Altner 1975, Kattmann 1977, Schaefer 1978a):

Exclusive Thinking	Inclusive Thinking
1. Selection; specialized thinking. Strict borders between disciplines.	1. Integrative thinking, multi-dimensional. No strict borders between disciplines.
2. **Maximal progress in a special field up to extreme efficiency. Economical thinking.**	2. **Moderated progress in a certain field by simultaneous consideration of other areas. Ecological thinking.**
3. **Decision thinking in the categories "Either - or", "First one thing - then the other" (see: switching the gear in a motor-car).**	3. **Thinking in compromises. Coexistence thinking in the categories "Both A and $\bar{A}$", "Here one thing - there at the same time the opposite one" (see: compartmenting in biosystems, e.g. cells).**

4. Objects are constructable and reproducible; they can be made "identical" (normed threads, spare parts). Identity thinking, ahistoric thinking. Demand of modern science and technology: Perfect reproducibility of things, perfect preciseness of description and explanation.	4. Objects are not constructable ad libitum. They possess individuality and never are "identical". Discriminative thinking; historical thinking. Unpreciseness in description and explanation of individual events. Lacking reproducibility of things.
5. Ego-centric development of technical products. Environment entirely under the view of human interests. Exploitation thinking, "wear and tear thinking".	5. Development of technical products with an altruistic view. Protection thinking, preservation thinking, nursery thinking (protection of animals, preservation of nature).
6. Demand for perfect functioning without any failure: Perfection thinking, deterministic thinking. Absolute security for the individual case (micro-stability).	6. Failures, lacking perfection are tolerated in individual cases on behalf of stability over a larger range (macro-stability, ultrastability). Flexibility thinking, adaptability thinking, evolutionary thinking, probabilistic thinking.
7. Model thinking. Empirical reality is viewed through the goggles of simplified models (atom $\rightarrow$ particle, organism $\rightarrow$ machine, ecosystem $\rightarrow$ energy source).	7. Object thinking. Cautious representation of empirical reality by models. Consideration of principal incorrectness of models (ecological equilibrium $\nrightarrow$ equilibrium of scales).

2.2. Training inclusive thinking by aspectizing subject matter.

It seems evident from this list that physics, chemistry, and technology are using more the exclusive kind of thinking, whereas biology stands and falls with the use of the inclusive one. In this situation many biology teachers are getting confused about how to adapt their instruction to the progress of biological sciences and how to train their students in "scientific thinking". Still part of them seek scientific thinking in the chemical or physical range, others more in the formal cybernetic range; i.e. they emphasize, more or less exclusively, o n e aspect of life. An increasing part of teachers, however, dare to open up their scope of aspects in biology teaching to problems of life in comprehensive, man-influenced ecosystems, including social, juridical, political, economical, esthetical, and also religious aspects. An instrument for such an aspectizing is shown in fig. 1. The lower aspects (grey zone) are more relevant for natural sciences, the upper aspects (white zone)

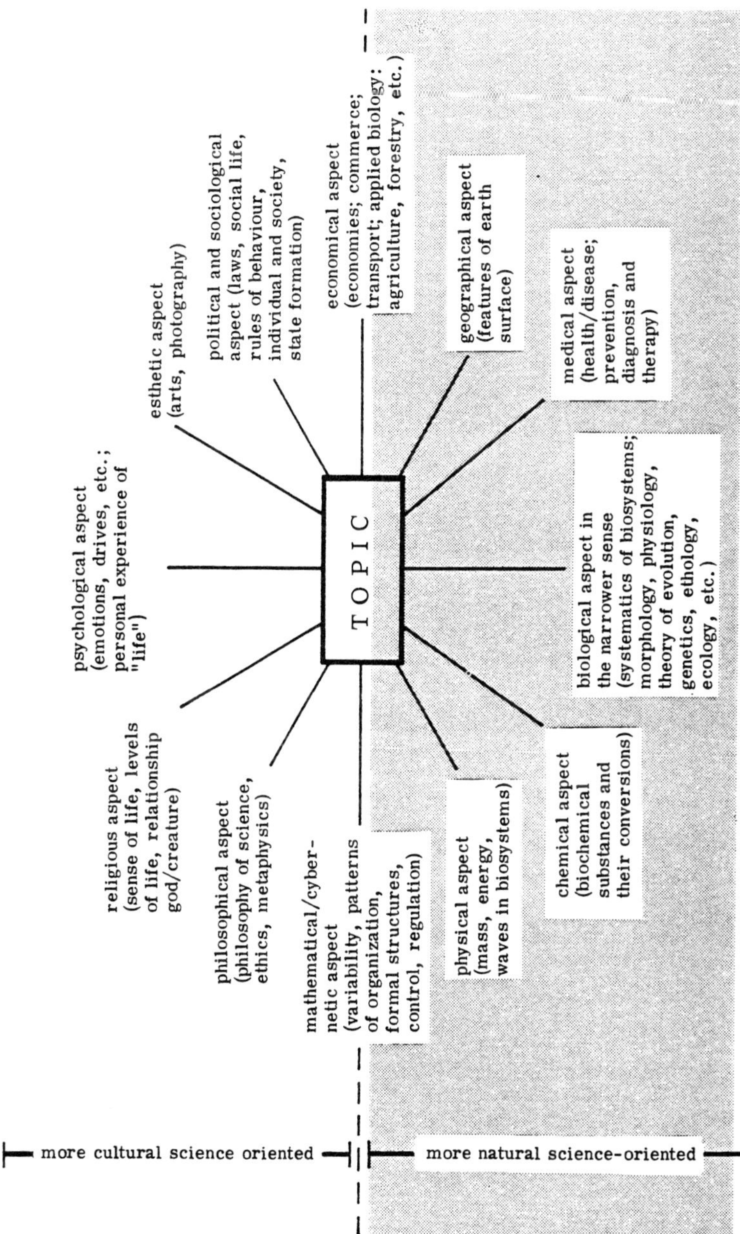

Fig. 1: 12 aspects of a topic which should be viewed in biology teaching, before reduction takes place to the narrower "biological" consideration. Pattern for aspectizing a phenomenon serving to develop and maintain an "inclusive thinking" (SCHAEFER 1975).

for social and cultural sciences. The mathematical/cybernetic and the economical aspects are bordercases; with respect to their methods they belong more to the upper zone, with respect to their objects and application fields more to the lower one.

An interesting approach to aspectizing subject matter has also recently been published by Bayrhuber 1978, in the field of health education. The reader of that article feels that "going round" the topic blood circulation the concept of circulation is enriched by a high variety of associations that makes it more colorful, more pleasant, and allows application in various situations of life.

Each biology teacher, when reflecting upon his own style of teaching, will find himself at home with one aspect or another. He will notice that, according to his pre-service training and the second subject he might have studied (chemistry, mathematics, arts), he will accentuate his instruction more from the chemical, the mathematical, the geographical, the esthetic or some other point of view. This one-sidedness is inevitable. The students, however, will always challenge the teacher from ever new aspects which are still included in their integral picture of the world, - aspects which the teacher will try to eliminate or neglect "for scientific reasons". The teacher normally prefers to proceed "exclusively", just like he was trained at the university, and he wants to offer a picture of the world based upon his own one or two teaching subjects.

This is not wrong per se, because also scientific biology of our universities is an essential constituent of our world, and the student should be prepared for it. However, there are many other areas of the same world (working place, family, free time hobby, political scenery), where scientific biology only appears in a context and not isolated, where - so to speak - life encounters us "in vivo" and not "in vitro".

In these areas, life phenomena like health, growth, irritability, sex, race, intelligence, learning, appear under manifold aspects and not only biologically. The student who has not learned in school to associate scientific view points with others, will find difficulties later on to integrate his knowledge into the context of life. This was one of the main reasons for the world-wide movement toward "integrated science" (see Frey and Häussler et al.).

Students should learn that there is certainly such a physical phenomenon like "energy", which can be reduced to and defined by the concepts of force and distance ("definire" means drawing borders, means exclusive thinking!); but they should also learn at school that in his private and public life it is much more the " context-energy " that counts (a term introduced by Schaefer et al.pointing out the integration of physics back into life); that means: for the citizen it is important to know whether 1 kJ olive oil energy or 1 kJ heating oil energy or 1 kJ heating oil energy or 1 kJ atomic energy or 1 kJ solar energy. Although in all cases the same quantity, the quality varies considerably according to the social context, the natural origin, the by-products, the after-effects, etc. (what manifests itself, for example, in a different price).

Thinking in contexts, "inclusive" thinking is an essential constituent of biological thinking. In this respect, i.e. in the degree of inclusive thinking, up to now biology, in particular human ecology, distinguishes themselves considerably from physics and chemistry. The biology teachers should realize the high demand society has upon their discipline. The physics and chemistry teachers, however, besides their important complementary role of training students more in exclusive thinking (i.e. in one kind of scientific preciseness), should also feel encouraged to open up their subject matter to more different points of view. The method of aspectizing is certainly not confined to biology teaching, although here the challenge to the teacher might be particularly strong. Physics and chemistry teachers can and should use more opportunity to accomplish their exclusive concepts to inclusive ones.

3. Inclusive Thinking with Inclusive Concepts

Thinking, in the sense of operational thinking after Piaget ("concrete" as well as "abstract"), is operating with concepts. Hence there are at least the following two ways of achieving a higher inclusivity of thinking: 1. The enlargement of the total number of concepts in memory from different fields, disciplines, aspects, and their interrelationships. That means: The extension of the whole epistemic structure of the learner to a larger variety of subject areas. (What in Germany is normally called "Allgemeinbildung"). 2. The development of single concepts with a high integrating power.

The first way would proceed via an enlargement of the total system of knowledge, the second one via an improvement of its elements. Of course, these two processes, as in all systems, are interrelated with each other: the total volume of knowledge will normally affect the quality of individual concepts (by the linkages between the concepts), and, reversely, the quality of particular concepts, especially those of a higher order of abstraction, the "advance organizers" after Ausubel 1960 or the "cognitive bridges" after Novak 1977) will influence the total character of knowledge.

In the following the second way will be illustrated. It will be shown, how both the interpersonal meaning (the "logic core") of a concept and the personal meaning (the "associative framework") can be investigated and influenced by teaching in order to develop "inclusive thinking". The investigations are basing upon the "burr model" of concepts (fig. 2).

3.1. The logic core of a concept. After an original suggestion by Dörner 1974 it was shown by Schaefer for the concept "ecosystem" and growth that the interpersonal meaning of a concept can be illustrated as the "logic core" in a diagramme containing the superordinate/subordinate relationships between concepts on the vertical and the system/part relationships on the horizontal axis, and the equivalent examples on a third axis (fig. 3). This kind of diagramme picks out the most significant and most frequent logic relationships we find in definitions

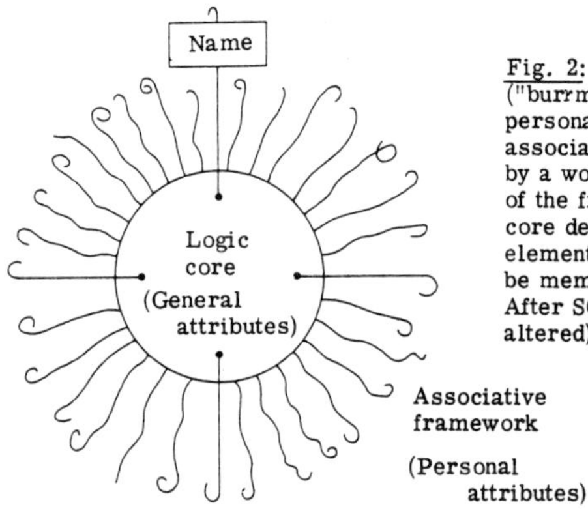

Fig. 2: Model of a concept ("burrmodel") showing the interpersonal logic core, the personal associations and the designation by a word. Some of the associations of the framework reach into the core demonstrating that also the elements of the logic core may be memorized by free association. After SCHAEFER 1979 (slightly altered).

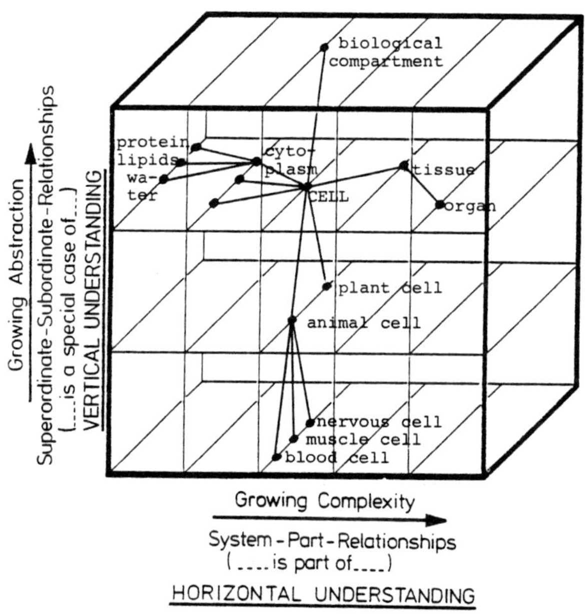

Fig. 3: Model of the "Logic Core" of a concept (here: cell), constructed from the two main logic relationships occurring in definitions (scheme after DÖRNER 1974)

about concepts given by test persons. It is applicable to objects resp. systems by using also the horizontal axis, as objects and systems consist of parts and hence we find system/part relationships here. However, if we examine only a property, a state, or a condition of an object or a system, we have to leave out the horizontal axis and only use the vertical together with the third (example) axis.

On the vertical axis we have particularly to take care of maintaining the category of concept analysis. If we have a concept representing a state variable, all other concepts below and above have to represent state variables, too. If the concept refers to a process, all other concepts in the vertical line have to refer to processes, too. The same holds true for material objects, etc. This is because the "examples" (concepts below) are always special cases of the concept under discussion and so belong to the same category.

Looking at fig. 3 it is obvious that concepts are becoming more inclusive the more they are either placed on the upper (abstract) or the right (complex) end of the diagramme themselves, or the more they are interlinked with such concepts of a "high order".

"Advance organizers", after Ausubel 1960, are defined in the vertical dimension, i.e. as superordinate concepts on a higher level of abstraction. According to empirical evidence (see Novak 1977) it may be taken as proved that such advance organizers are able to facilitate structuring new experience and to improve the learning capacity of a person. In this function they turn out to be "inclusive", as they connect different areas of knowledge with each other. Cybernetic concepts, for example, are concepts of this type, as was demonstrated by Hassenstein 1967, Kuhn 1967, Schaefer 1972b, 1976, Eulefeld and Schaefer 1974, Bayrhuber and Schaefer 1978).

Another possible kind of inclusive concepts which, however, empirical researchers have not yet paid equal attention to so far are concepts of the highly complex category, which are placed in fig. 3 on the very right end. These concepts, like "ecosystem", "biosphere", "world", or whatever always are designating objects of a high degree of complexity, i.e. consisting of a large number of elements and their interrelations. If such concepts would be taught at school with accuracy, a high number of different views and different areas of knowledge would be connected with each other. In other words: If we knew that different things are part of the same object or system and we knew the interrelations within the system, the concepts referring to the parts would become inclusive; unconnected facts would become structured, "rote learning" would change to "meaningful learning". However, the "meaning" here would be represented by system-part relationships and would not, as in many publications on concept formation, be restricted to superordinate-subordinate relationships.

Empirical investigations upon the latter have been frequently carried out (see above). Research upon the effects of deliberate training in highly complex systems on the development of inclusive concepts, however, is rather rare (see Dörner et al.). New investigations of this kind, concentrating upon the concepts "ecosystem" and "environment" have been initiated by the author one year ago. First results upon the concept "ecosystem" will be reported in chapter 4.

3.2. The associative framework of concepts. In order to find out the "personal meaning" of a word for a test person, the method of free associations has sometimes been used. Although there is not yet any rounded-up psychological theory about associations to work with (see discussions by Deese 1965, Landauer 1978), the author was able to show interesting regularities in the percentage of associations in certain categories (see investigations on concepts like Equilibrium, Growth, Behaviour, Stimulus, Health, Individual, and others; Schaefer since 1972, e.g. Eulefeld and Schaefer 1974). Some of the main results of these investigations are here summarized in brief:

1. It was again proved that in the spectrum of associations round a biological term there are plenty of associations coming from quite different, also non-biological fields, and that quite a number of associations are surprisingly far away from the scientific meaning of the concept.

2. It was found that in spite of rather different personal backgrounds of the test persons (what was demonstrated by the specifity of individual associations) the groups of associations are rather stable in a certain population. That means: Independent from the kind of individual associations which, of course, change from week to week, from person to person according to everyday experience the percentage of associations falling into a certain category is rather constant. Obviously there is a tendency to associate into certain directions which is characteristic for a population in respect to a specific concept. These tendencies have to be regarded as an effective attribute of the concept itself and should be object of educational planning equally to the logic (scientific) core of the concept.

3. There is empirical evidence for the general hypothesis that the percentages of free associations in certain categories and the percentages of definition elements falling into the same categories are somehow corresponding to each other. Although in single cases (especially in categories with a particular emotional load) both percentages may deviate from each other, normally the maxima and minima in the distribution of free associations and free definitions are alike (see figure 4 upon "ecosystem"). This proves that free definitions given for a particular concept are highly influenced from a subliminal subliminal associational level. This influence has to be taken into serious consideration in educational planning, as it might be responsible for the failure of teaching "pure" scientific concepts in schools.

These previous results, although still far from explanation by one of the classical psychological theories, deserve some attention by teachers and curriculum planners. This will be demonstrated in the following with the concept "ecosystem", which is one of the most important concepts of environmental education to-day and needs special training in "inclusive thinking".

4. "Ecosystem" as an Inclusive Concept

4.1. The logic core, as it should be. Fig. 6 demonstrates the pattern of logic relationships identifying the concept ecosystem. It is, in its central part, a representation of the interpersonal (international, scientific) meaning of the word "ecosystem", i.e. a model of the logic core of the concept.

An adequate definition of an ecosystem, according to fig. 5, would be:

An ecosystem is a sub-system of our biosphere composed of producers, consumers, decomposers, and abiotic factors, like a forest, a lake, a moor; it is able to maintain an ecological equilibrium (fluctuations between certains limits) over a certain period by means of self-reproduction of sub-units and self-regulation of the whole.

Although there is still some discussion among ecologists about the phenomenon of self-regulation being an essential criterion of an ecosystem or not, most ecologists to-day agree upon this criterion (see Ellenberg 1973, Odum 1976). So the above definition can be taken as an agreement among the scientific community of ecologists, and it can be used as a measure for comprehension in a research project on concept formation.

Fig. 6 demonstrates, how the range-specific understanding of a particular ecosystem, say a lake, can be extended to a general understanding of ecosystems by comparing a lake with other examples of ecosystems. This understanding, however, although more general, is still rather "range-specific" in respect to ecology. So the comparison of ecosystems with other, non-ecological self-regulating systems will yield a better understanding of the concept ecosystem on the next higher abstraction level, i.e. on the level of the concept "equilibrium system" (regulation system). On this level it turns out that there are two distinct kinds of equilibrium systems: those with special information channels (homeostatic systems with indirect feed-back via "messages", like nervous systems and most technical systems), and those without (systems with direct interaction of their parts, like chemical buffers, ecosystems). If ecosystems are misinterpreted as homeostatic systems, researchers will in vain seek for information channels, set points, regulators, etc. So the knowledge of the two different kinds of regulation system necessarily improves understanding of the concept "ecosystem". This kind of understanding by more generalized, abstract concepts is called by Greeno (1977) "explicit" and by Schaefer (1979) "vertical" understanding (the latter according to the geometrical setup of fig. 3, which is based upon the normal versions of "higher" and "lower" abstraction level).

"Equilibrium System", "Cyclic Dynamic System", "Dynamic System", and finally "System" therefore are continous steps on the way to higher abstraction and to a higher degree of understanding of what is an ecosystem. They are "advance organizers" in the sense of Ausubel.

These abstract concepts of systems theory, however, can only be really effective in different fields of experience, if there is enough basic empirical knowledge in these fields which has been connected with them. So the concept of "regulation cycle" proved to structure new ecological experience only if the linkage between ecology and cybernetics had been sufficiently trained before in the curriculum (Eulefeld and Schaefer 1974, English translation 1978, p. 50). This again gives a hint upon the marked influence of associations on thinking. The concept of "regulation system" is able to link knowledge from physiology, ecology, chemistry, physics/technology, economics, sociology, even psychology together, as all these fields contain examples of regulation that can be subsumed under this superordinate concept, - provided the subsumption has been prepared by bringing those fields together in the associations of the students.

One way of training inclusive thinking is therefore the training of superordinate concepts, of "subsumers" (Novak 1977) together with various examples from different subject areas. Textbooks for biology at schools and universities deal to a large extent with system/part relationships of biosystems (morphology, anatomy, physiology, genetics, etc.); superordinate/subordinate relationships are specified only on a rather low level of abstraction (see lower part of fig. 3). Sections on systems theory are lacking or rare. So the "vertical understanding" of the students normally is little developed, especially when there is no obligation of studying biology at the university together with mathematics. The author therefore, in the last ten years, has published instructional devices under the head-line "cybernetic biology" in order to help the teacher to include more abstract levels into biology teaching (Schaefer 1972a, Eulefeld and Schaefer 1974. - English translation 1978 -, Bayrhuber and Schaefer 1978, Dylla and Schaefer 1978). The evaluation of these curriculum units has proved the effect of cybernetic thinking on vertical understanding in biology and on the development of inclusive thinking by a rich pattern of associations to different subject areas.

4.2. The logic core, as it is

The analysis of definitions given by students in the age of 13/14 and 17/18 and by biology teachers yields an interesting insight into the epistemic structure of the test persons in regard to the concept "ecosystem". There were four different kinds of analyses carried out:
1. The qualitative nature of the definition in respect to right, wrong, etc., 2. The determination of the definition volume in the Dörner scheme (more above, more below, more right, more left,

etc.). 4. The distribution of different content areas in the definition, compared with the categories found in free associations.

Previous results taken from schools in Northern Germany (Hamburg, Kiel and surroundings) are summarized in the following:

1. In the age of 13/14 about 75 % of the students are not able to give a correct or at least incomplete definition of an ecosystem. 41 % of the students gave definitions which were of the associative kind and not logically structured. About 22 % found a partly correct definition, which however was incomplete. None of the definition was fully correct, but 5 % were very close to correctness.
In the age of 17/18 the situation has markedly changed: Only 22 % were not able to give a correct or at least incomplete answer. (It is surprising that the result of the tested biology teachers, 16 %, is not very far from this!). The number of associative definitions has dropped down to 17 %, whereas the amount of incomplete definitions has increased up to 41 %. In this age group we find about 18 % absolutely correct definitions (what again is not far from the result 23 %, gained by teachers).

2. In the horizontal dimension of the Logic Core we find a shift from the maximum being H 1 in the 13/14 age group (55 %) towards the maximum being H 2 in the 17/18 age group (59 %), which is also maintained by the teachers with 53 %. Furthermore, we find a marked increase of H 3: from 0 % in 13/14 to 7 % in 17/18, and to 10 % in teachers. This means that with growing education the number of system-part-levels and so the degree of inclusive thinking in the horizontal dimension seems to rise.

3. In the vertical dimension of the Logic Core the maximum of the 13/14 year olds is in V 2 (two levels of abstraction) with 34 % of all students. This maximum rises up to V 3 in the 17/18 year olds group (38 %). Again it is also maintained by the teachers (40 %), but the teachers distinguish themselves from the students by a rather high quote of >V 3 (33 %), what shows that with higher education the ability to move through different levels of abstraction grows visibly. This speaks for a growing ability for inclusive thinking in the vertical dimension in higher stages of education, an assumption which is also supported by point 6.

4. The number of examples in the definitions (examples in the strict sense as subordinated concepts) jumps from the maximum in E 1 (33 % of the 13/14 year olds) to the maximum in E 3 (38 % of the 17/18 year olds) and further increases to the maximum in >E 3 (30 % of the teachers). This is surely to be seen in connection with the increase of the number of levels in the vertical dimension reported above, which leads inevitably to a higher rate of examples, i.e. to a broader spectrum of views, applications, specifications. It is symptomatic for inclusive thinking.

5. The position of definitions within the Dörner scheme spreads in the horizontal dimension from medium to the right end and left end.

That means: More complex systems as well as more details (elements) of systems can be found in the definitions. This is obviously an effect of the biological system-part training prevailing in school and university curricula.

6. In all age groups the emphasis of the definition in the vertical dimension lies in the upper (more abstract) region. Definitions mostly move on this level by using superordinate terms like "system", "process", "science", etc. and then tend to remain on this level. There is an increase again through the ages from 28 % to 49 % to 63 % and a growing differentiation towards the upper level in the teachers. This shows that a growing number of high-order abstract terms is available in memory in the higher age groups, which may function as advance organizers in inclusive thinking.

7. As fig. 4 demonstrates, the distribution of content areas within the definitions is very similar to that of the free associations to the key-word ecosystem. So this distribution will be analyzed in the following chapter dealing with associations. One striking result gained from the definition tests should, however, be pointed out already here: The distribution does not suit to the "standard definition" of an ecosystem (see fig. 5). It contains quite correct maxima in the biological and also formal field, but it is lacking equivalent values in the range of non-living nature (which is part of ecosystems!), economy (which is also an essential factor influencing ecosystems and being dependent from their status), and environmental affairs (which cannot, although having a political side, be detached from the scientific concept of an ecosystem).

Points 1 to 7 of these investigations, carried out in West-Germany, correspond to similar results gained in the Philippines in Feb. 1979. These results will be published in a cross-cultural study on concept formation at a later stage.

4.3. The associative framework, as it should be. A hopeful approach to inclusive thinking can certainly be the interlinkage of associations from different subject areas on a non-logic, non-systematic level. People will use advance organizers and logic relationships as long as they are moving in a "logic world" (e.g. world of science, world of technology). But there are plenty of examples showing that people forget about scientific, logic relationships and fall back to associative behaviour when moving in a "non-logic" (that means: non-systematic, not purposely structured) world which is predominant in parts of our daily life (family, friendship, sports, hobby etc.). Preparing students for life and not only for specific professions (the major aim of primary and secondary school teaching!) would mean then concept formation towards inclusive thinking from both ends, i.e. for both types of life situations, the "logic" and the "non-logic" ones.

If we are going to make a concept associable to all fields of everyday life where it may be important, we have to take care that its associative framework covers all these fields with an adequate degree.

The concept ecosystem seems important in at least the following areas of consideration: 1. Living Nature (Biology, Ecology); 2. Non-living Nature (Physics, Chemistry, Metrology, Geology, etc.); 3. Environmental Affairs (Environmental Policy, Pollution, Preservation of Nature etc.); 4. Agriculture, Horticulture; 5. Science (Terminology, Research, University, etc.); 6. Formal disciplines (Mathematics, Systems Theory); 7. Economies (Industry, Commerce, Resources, Structure of Society, etc.); 8. School (Teaching Events in connection with Ecosystem). So we have to expect a spectrum of associations round the term "ecosystem" which contains to a certain degree associations from all these eight categories. In the following chapter empirical investigations will throw a light upon this theoretical expectation.

4.4. The associative framework, as it is. In the last ten years, investigations on free associations and free definitions of various biological concepts have been carried out in West-Germany. The investigations have been extended also to Israel and to the Philippines. The concept "ecosystem" was studied since one year in different West-German schools and age groups. The total number of tested students in the age group 13/14 years is up to now 206, in the age group 17/18 years 163, and of biology teachers 52. There is a rather clear picture already about the distribution of associations and definitions upon the concept "ecosystem" in these groups, which is illustrated by one selected sample of classes in fig. 4.

We find in the hitherto examined West-German and also Philippine populations the following results:

1. There is a marked correlation between the distribution of free associations and the distribution of free definition elements. The correlation coefficient in the total population examined up to now is between + 0.7 and + 0.8.
2. The distribution of free associations does not cover all eight expected categories to a satisfying extent.
3. There is a clear deficiency in the associative framework of grown-ups (17/18 year olds and biology teachers, cf. fig. 4, right) with respect to agriculture, economy, technology. It is surprising to see that the younger students (13/14 year olds, fig. 4, left), although to a great extent not yet knowing what an ecosystem "is" (see definitions, category "Unknown"), associate with this word much more elements from agriculture, economy, and technology/physics. It seems, as if school and university, by training a narrow scientific concept of ecosystem, "spoil" the associative framework of this word.
4. In all examined cases the associative framework of the test persons contains two distinct maxima: one in the category "Living Nature", the other in the category "Formal and Abstract Terms".

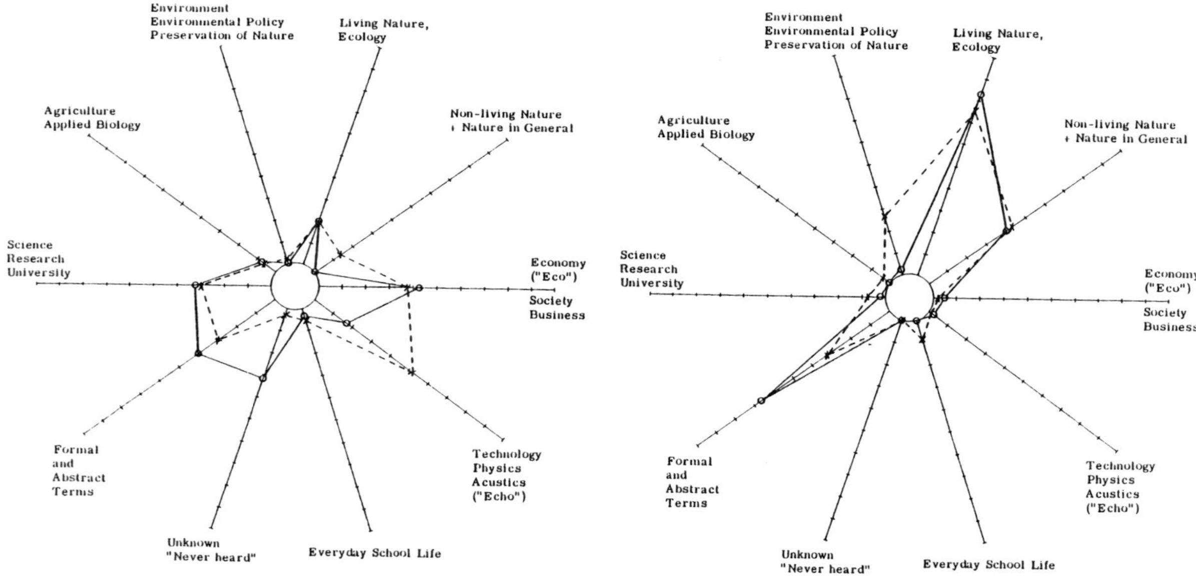

Fig. 4.: Distribution of Free Associations (---) and elements of Free Definitions (——) upon the key-word ECOSYSTEM. Gymnasium Hamburg. One mark means 5 %.

Left: 12/13 year olds, before instruction on "ecosystem"; n = 162

Right: 17/18 year olds, before special course in ecology; n = 148

Whereas the first one is not surprising, the latter was unexpected, especially in the younger age group (12/13, 14/15). The individual associations and definitions show that this maximum is induced by the key-word "system" which seems to provoke a great deal of abstract terms.

These results may cause some reflections among teachers about the type of environmental education they are practising in school. The deficiencies pointed out in 3 should make them think about including more concrete examples and tasks from the fields of *applied* ecology into teaching, if they really want to develop an "inclusive" concept of an ecosystem.

5. Logic Core - Associative Framework - Action

In this paper a proposal has been made to develop inclusive thinking by means of concepts. The whole decision tree for pedagogical devices may be pictured like this:

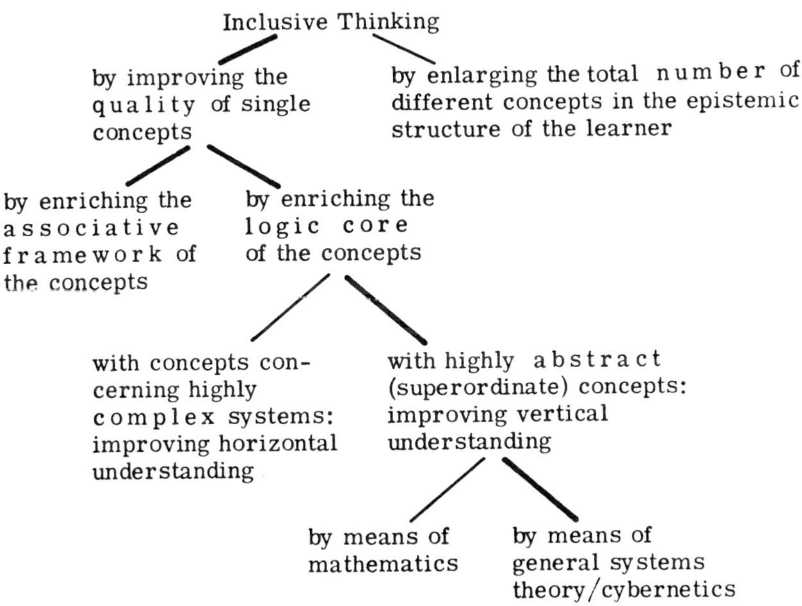

The way marked by thick lines has been worked out in the last ten years by the author and his co-workers. It proved to be successful (literature see above). As long as we regard concepts to have an affective component included in the associative framework influencing the actions of a person, we avoid the mistake of pure cognitive view of concept training. Interviews with 20 students whose definitions and associations upon "ecosystem" had been examined, showed that the cognitive training of the logic core had an effect on students' actions (as anticipated by them) only through the "filter" of the associative framework. The logic core of the concept "ecosystem" was memorized and used in such situations where it was a s s o c i a t e d with the situation in some way. Otherwise it was not remembered at all. Furthermore, it turned out that emotions connected with the associations were determining to a great extent the anticipated actions. So the knowledge of the word "system" and the pure understanding of its scientific meaning did not allow any prediction upon what the student would do with systems in practical life. But the analysis of free associations revealed two different types of reaction to the word: one student associated with "system" the terms "entity" and "life". Asked about his possible future actions in respect to systems, he answered in the sense of "positive engagement for scientific ecology". That means: The concept "system" (within ecosystem) was associated with positive feelings, which brought about pleasant associations (or vice versa) and could be recognized by them. So a prediction in the direction "positive engagement ..." was possible and was later confirmed in the interview.

The opposite was true with another student, who associated "mathematics", "abstraction", "dead". The prediction concluded from his associations was: negative attitude towards systems, general antipathy against formalization. The interview again confirmed this expectation: The student rejected any systematic analysis of nature in the way how science does; he was a strong opponent of science and promised "never to become a scientist". So the free associations seem to allow some assessment of future actions on the emotional level. The findings by the method of interview have, of course, to be confirmed by the study of real actions in life instead of only anticipated ones. These studies are being carried out. Some of the data presently available have been illustrated in fig. 5. This figure gives a picture of possible relationships between the logic core of "ecosystem" and the field of everyday actions. It also shows the high significance of free associations with respect to the assessment of future impulses originating from concepts. One highly important impulse found in the interviews was: reflection upon the epistemic structure in oneself. Any controversion in actions or feelings may lead back (see reflected arrow in fig. 5) to the logic core and clarify the conceptual situation before acting. This kind of "internalized action" receives growing importance in our complicated and controversial pluralistic world.

Further research and curriculum development have to prove to what degree concepts like "ecosystem" trained in an inclusive way are able to determine man's actions in everyday a n d scientific life.

The author dares the expectation that by this kind of training it will be possible to develop a new "ecology of mind" (Bateson 1972), which would continue the present ecological movement to an "inside ecology".

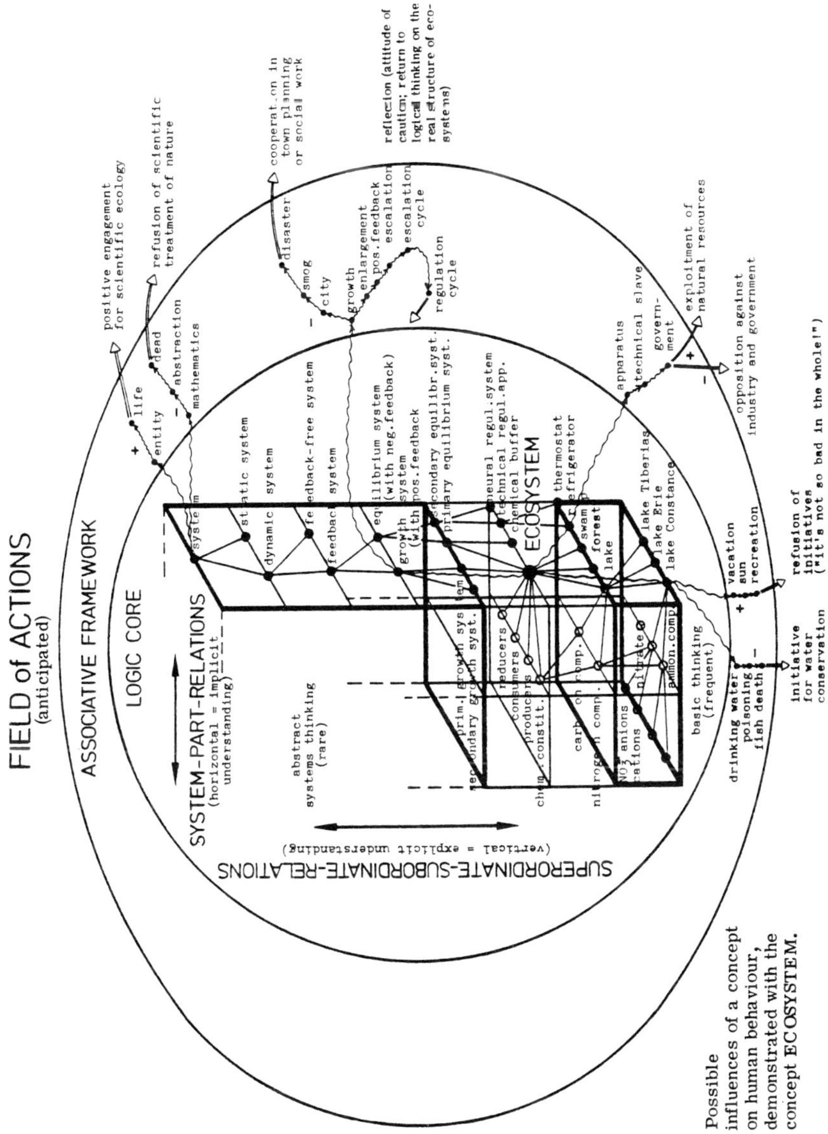

Fig. 5: Possible influences of a concept on human behaviour, demonstrated with the concept **ECOSYSTEM.**

6. Literature

ALTNER, G., 1975: Die Erfahrung des Lebendigen durch die moderne Biologie. epd Dokumentation 31, 35-36.

ASHBY, R., 1956: An introduction to cybernetics. Chapman + Hall Ltd.: London.

AUSUBEL, D.P., 1960: The Use of Advance Organizers in the Learning and Retention of Meaningful Verbal Material. J. Educ. Psych. 51, 267-272.

BATESON, H., 1972: Steps to an Ecology of Mind. Chandler Publ. Comp., Ballentine Books, Random House: New York, Toronto.

BAYRHUBER, H., 1978: Aspekte des Themas Blutkreislauf. Weltgesundheitstag 1978. Bundesvereinigung für Gesundheitserziehung e.V.: Bonn.

BAYRHUBER, H. and SCHAEFER, G., 1978: Kybernetische Biologie. Curriculum Unit (half-year's course) for the age 17 to 19. Aulis: Köln.

CARROLL, J.B., 1964: Language and Thought. Englewood Cliffs: New Jersey.

DEESE, J., 1965: The Structure of Associations in Language and Thought. The John Hopkin's Press: Baltimore.

DÖRNER, D., 1974: Kognitive Struktur und Information. Paper presented to the workshop "Struktur und Information als Determinanten von Lehr- und Lernprozessen", organized by the German Society for Cybernetics (DGK), Bad Homburg.

DÖRNER, D., 1976: Psychologisches Experiment: Wie Menschen eine Welt verbessern wollten und sie dabei zerstörten. Bild der Wiss. 1976, 48-53.

DÖRNER, D., 1979: Mängel menschlichen Denkens beim Umgang mit sehr komplexen Systemen. In "Ökologische Außenstelle" (ed.): Berichte der ökologischen Außenstelle Schlüchtern der Johann-Wolfgang-Goethe-Universität Frankfurt/M., Heft 2, 43-62.

DYLLA, K. and SCHAEFER, G., 1978: Tiere sind anders. Curriculum Unit for the age 8 to 10. Aulis: Köln.

ELLENBERG, H. (ed.), 1973: Ökosystemforschung. Springer: Heidelberg.

EULEFELD, C. and SCHAEFER, G., 1974: Biologisches Gleichgewicht. Curriculum Unit for the age 12 to 14. Aulis: Köln (English translation 1978).

FREY, K. and HÄUSSLER, P. (ed.), 1973: Integriertes Curriculum Naturwissenschaft: Theoretische Grundlagen und Ansätze. 4. IPN Symposion. Beltz: Weinheim, Basel.

GREENO, H.G., 1977: Analysis of understanding in problem solving. Paper presented to the workshop "Developmental Models of Thinking", IPN: Kiel.

HASSENSTEIN, B., 1967: Biologische Kybernetik. Quelle & Meyer: Heidelberg.

KALLEN, D., 1978: Ad hoc conference on society and the study of science, mathematics and technology. Standing Conference of

European Ministers of Education, Strasbourg.
KATTMANN, U., 1977: Bezugspunkt Mensch. Aulis: Köln.
KUHN, D.J., 1967: A Study of Varying Modes of Topical Presentation in Elementary College Biology to Determine the Effect of Advance Organizers in Knowledge. Ph.D. thesis, Purdue Univ., USA.
LANDAUER, T.K., 1978: Quantity and Quality in Knowledge. In Scandura, J.M. and Brainerd, C.J. (ed.): Structural/Process Models of Complex Human Behaviour. Sijthoff + Noordhoff: Alphen aan den Rijn (Netherlands), 177-206.
NOVAK, J.D., 1977: A Theory of Education. Cornell University Press: Ithaca and London.
ODUM, H.T. and E.C., 1976: Energy Basis for Man and Nature. Mc Graw Hill: New York etc.
SCHAEFER, G., 1972a: Kybernetik und Biologie. Metzler: Stuttgart.
SCHAEFER, G., 1972b: Kybernetische Konzepte als Instrumente für die Begriffsbildung in der Biologie. Unpublished manuscript, available at IPN Kiel.
SCHAEFER, G., 1973: Kybernetische Zeichensprache als Hilfsmittel zur Strukturierung eines integrierten Curriculum Naturwissenschaft. In Frey, K. und Häußler, P. (see above), 185-203.
SCHAEFER, G., 1975: Ökologie - Lehrfach oder Unterrichtsprinzip? In Müller, P. (ed.): Verhandl.d.Ges.f.Ökologie Erlangen, Junk: Den Haag, 269-274.
SCHAEFER, G., 1976: Was ist Wachstum? Eine empirische Untersuchung zur Begriffsbildung. In Schaefer, G. et al. (ed.): Wachsende Systeme. Leitthemen 1/76. Westermann: Braunschweig, (now Aulis: Köln), 58-89.
SCHAEFER, G., 1977: "Energetik" - Kybernetik. In Wenk, K. and Trommer, G. (ed.): Naturerscheinung Energie. Leitthemen 1/77. Westermann: Braunschweig (now Aulis: Köln), 204-216.
SCHAEFER, G., 1978a: Inklusives Denken - Leitlinie für den Unterricht. In Trommer, G. and Wenk, K. (ed.): Leben in Ökosystemen. Leitthemen 1/78. Westermann: Braunschweig (now Aulis: Köln), 10-29.
SCHAEFER, G., 1978b: Kybernetik im Biologieunterricht. In Schaefer, G. (ed.): Biokybernetik. Unterricht Biologie Heft 21, Mai 1978, 2-10.
SCHAEFER, G., 1979: Concept Formation in Biology: The Concept "Growth". Eur.J.Sci.Educ. vol. 1, No. 1, 87-101.
SEIFERT, A., 1977: Über die Verwendbarkeit des Begriffes "Gesundheit" in der Wissenschaftssprache der Biologie. Thesis for State's Examination, Univ.of Kiel (W.-Germany). Available as copy at IPN Kiel.

TEACHING FOR ACTION BEYOND SCHOOL: THEORETICAL AND PRACTICAL PROBLEMS

by C. Sutton and D. Tomley

Is the Theory adequate?

In this meeting we are dealing with an important theoretical idea. It is that connected knowledge is transferrable knowledge, accessible to its possessor from many different starting points and at many different times, and therefore usable in contexts different from those in which it was acquired. For school biology, just as much as for any other subject, it suggests that if important concepts are to be applied in private and professional life, things done at school will have to be seen by the learner to link with everyday life. We are concerned with the nature of those links and how they are formed, and also with the design of teaching methods which best assist their formation.

Many of us are already convinced of the importance of making connections to out-of-school knowledge. But is it really established that richness of such connections does in fact lead to the use of concepts in later life? Does that have to be tested empirically, and if so how exactly would it be done? Or is it simply logically true, inherent in the meaning of what we call knowledge? What do we learn from all the empirical studies of transfer that experimental psychologists have attempted? Are similar studies necessary now to see what happens when biology teachers deliberately 'teach for transfer' on biosocial topics? Before we got too deeply into the lines of argument developed by Schaefer[1] and Kelly[2], some scepticism is perhaps appropriate. What would one say to teachers for whom it is equally "self-evident" that the way to develop 'good thinking ability' is to isolate the child from his everyday experience, to cut him off from the everyday meanings of words, and help him develop a coherent understanding of the language of the discipline? Such teachers might argue that you have to acquire the mental tools first and apply them later, and that you have to specialise before you can integrate. To mix discussion of biological 'growth' with chat about the growth of cities, and bank accounts, is to risk losing all precision in thought[3]. And they might suggest that Kelly's "issue studies" are best left to the current affairs teachers, leaving the science staff free to consolidate their "knowledge studies".

Whether or not they require empirical test, the resolution of some of these questions will certainly demand much more detailed articulation of the theoretical position. What exactly is meant by connectedness when we talk of 'connected knowledge'? Have we got a means of monitoring its development? Can we spell out more fully the procedures in the classroom that are expected to bring it about, and so on? Papers such as those of Schaefer[1,5] have pushed forward the clarification of the theory considerably, and in the first part of this one we shall attempt a further statement of the theoretical position, drawing on British interests in the role of language in learning. We shall also discuss why these ideas have come to prominence recently.

The essential proposition is that human beings are continuously involved in sense-making mental activity, exploring for meaning everything new that they encounter. Meaning arises out of the ways in which they connect this new experience with what they already know, and it is therefore indefinitely expandable as more and more connections are made. Meaning is not something fixed which can be 'given' to a learner complete in itself, but is changed, enriched, and recorderd as time goes on. If that seems "only common sense" we should notice how strongly it conflicts with the desire that has animated science since the seventeenth century, to introduce precision into language by fixing, or trying to fix, the meanings of words, confining them within precise definitions:

> "When our knowledge becomes perfectly exact and purely intellectual, we require a language which shall also be exact and intellectual; we shall exclude alike vagueness and fancy, imperfection and superfluity, in which each term shall convey a meaning steadily fixed and rigourously limited. Such is the language of science."[6]

As a result, the members of scientific community are in danger of forgetting that the meaning of any word or idea has to be built in the mind of the learner. The practice of giving more emphasis to dictation of definitions rather than to discussion of meanings is one symptom of this forgetfulness. (So is the tendency to value more the pupil's ability to repeat the dictated definition than his skill in illustrating an understanding of it.)

Schaefer suggests that it is not just the logical core of a concept - its denotative meaning - which matters. The multiple chance connections in its "associative framework" - i.e. its connotative meaning (or, as it has been called, the "penumbra of meaning") matter too. They are the means, it is being asserted, by which an idea becomes capable of influencing a person's behaviour in a variety of situations, including those he or she meets as a citizen outside and beyond school.

Is that idea really true?

Is it really usable as a means of devising teaching/learning strategies?

We think that it is true, and that it is usable, provided that rapid techniques can be found for monitoring successive changes in a learner's private version of a particular concept. If science teachers are to take greater account of the idiosyncratic features of each learner's private concepts (though not at the expense of developing precision in the public version), they will need much better means than we now have for discerning and describing these features.

Some reasons for the growth in interest in the content of the learner's thought: Before considering the difficulty of monitoring the content of someone else's thought, it is interesting to consider how and why interest in doing so has grown recently. Several influences seem to have been at work. One is the decline of strict behaviourism. It has become respectable to speculate on the unseen intangible world of what goes on inside a learner's head. As scientists we should perhaps maintain some check on the products of such speculation, but many people are now more ready to recognize that science always involves an interplay of imagination and observation, and (with Kekule) we need no longer be afraid to dream, as long as we know that we are going to test those dreams.

...And let's face it, "concept burrs" and "cognitive structures" are dreams - creations of our thought and talk, rather than something we shall see by looking into someone's skull. Mental activity is itself a construct, inferred from behaviour. Some years ago people were for that reason reluctant to talk about it. Now more people are talking about it. Some are doing so at the 'purely' theoretical level. Others, (under the influence of neurophysiology and/or computer models of artificial intelligence) hope that speculations about "connected ideas" will sooner or later have a relationship with observable connections, and events in brain anatomy and physiology.

Another influence comes from tape recording classroom discussions. In Britain, some teachers have come to value the process of making multiple connections to old knowledge after studying how school pupils discuss, talk about, and talk out problematic and arguable situations. Groups led by James Britton and Douglas Barnes have examined the details of discussion in small groups. They come to a view which is nicely summarized in Barnes' distinction between 'school knowledge' and 'action knowledge': [7]

> "School knowledge is the knowledge which someone else presents to us. We partly grasp it, enough to answer the teacher's questions, to do exercises, or to answer examination questions, but it remains someone else's knowledge, not ours. If we never use this knowledge we probably forget it. In so far as we use knowledge for our own purposes however we begin to incorporate it into our view of the world, and to use knowledge becomes incorporated into that view of the world on which our actions are based I would say that it has become 'action knowledge'."
> "...An important task for the learner in school is that of converting school knowledge into action knowledge."

Other people have developed similar ideas from research more directly in cognitive psychology. For example, Donaldson[8], in her critique of Piaget, argues that children she and her co-workers have observed, cope well when they see the problem they are working on as "embedded" in the totality of their understanding about what human beings do. They reason better when their problem is linked in with both cognitive and affective aspects of their understanding of logic and of people.

Sociologists have long been aware of the compartmentalisation of the curriculum, and the sharp separation between what teachers and pupils take for granted as allowable knowledge and/or activity in a particular school setting. In Bernstein's terminology[9], the traditional secondary school curriculum is characterized by strong "classification" and "framing". If the theory of meaning that we are using is correct, then it would seem that biology teaching must undergo a loosening of the "framing", allowing pupil and teacher more freedom in the choice of lesson activities. On the other hand, since those demarcations were probably socially functional, they won't disappear just because some curriculum developer says they should. Even if overwhelming evidence were obtained that compartmentalisation retards application of learning (and, as we have said, that is still unproved), any realistic appraisal of how the curriculum changes over time would still have to understand the forces which maintain disconnectedness.

Another influence in the shift towards a greater interest in the structure of pupils' knowledge might be described as more to do with changes in the dominant epistemological views of the time, rather than changes in theories about learning. Current views of what knowledge is, make it less like mental furniture and more like mental spectacles, less like a library that we accumulate and more like a set of mental tools, whose quality, and the way in which they are used, will determine how we make sense of future experience. This shift arises from

(i) studies of perception which show that existing knowledge influences how we interpret new experiences, and
(ii) historical studies which show that people get 'locked in' to an existing theoretical framework.

After repeated attempts to describe science as both a body of useful knowledge and a set of useful mental skills, and a couple of decades recently when the 'process' school has been very strong, the pendulum now seems to be swinging again in favour of the conceptual view. Science, in Novak's phrase, is seen as "evolving populations of concepts"[10].

Scientific processes are possibly just an outcome of having certain concepts (e.g. "Designing a controlled experiment" could be a procedure made possible by holding the concept of a "variable").
If that is correct, it is a further reason why we urgently need techniques and notations for describing a child's existing concepts and their interrelationships.

Practical Problems

That school knowledge can enlighten out-of-school experiences, and that the latter can broaden the former, we ourselves no longer question. But what can we do to make it more likely that there is this reciprocity? How can we ensure that it is manifest to the learners themselves, and to more teachers?

It seems that there are at least three key areas:

(i) First we need to be able to find out what the pupils know of a given concept, i.e. to explore the private concepts available to them, including their non-logical aspects.
(ii) We schould try to trace pupils in the act of expanding and changing particular concepts.

(iii) Ultimately it would still be desirable to do a longitudinal study and follow these pupils into life after school.

The problem of the diagnostic technique is formidable. Several methods have been used to try to find out what sort of structure of ideas learners already possess. (For a summary see Sutton[11]). They include clinical interviews; word association tasks; asking learners to write definitions or to choose a preferred one from among several correct ones, and tasks which involve bipolar dimensions on which an idea is rated. Some of these techniques are far too time-consuming for a teacher to use, as he attempts to 'get inside the learner's head', and to understand how he or she sees the world. Also some of them do not adequately allow for the fluidity of human thought. Yet if all this theory is to be turned to practical use in the classroom, a teacher needs relatively quick techniques for diagnosing cognitive structure.

Ausubel's theories have been publicised in terms of 'advance organizers', but according to one of his principal interpreters[12], the more important feature of those theories is the emphasis on individual differences in each learner's cognitive structure. Not only are his private concepts all variants of the publicly agreed version; so is his overall cognitive map: the way in which these concepts are interrelated. The attempt to describe them, and to identify 'subsumers' under which new knowledge can be incorporated is crucial.

Over the past two years, some Leicestershire teachers have undertaken enquiries in areas relevant to this present discussion. R.A. Fisher studied how student hairdressers interpreted various problems involving moisture in the air, rates of drying of hair, etc. He tape-recorded their interpretation of problem situations, and found that some had concepts of evaporation and condensation which were effectively linked to a particulate theory, and others at least did not use such a mental connection even if they had it. He found that one of several word association and word sorting tasks seemed to correlate fairly well with the results of these problem interpretation exercises.[13] Teaching strategies appropriate to these different patterns of existing thought (and therefore different available subsumption systems) could then be devised.

J.C. Horsfield tried to examine the process of expansion of pupils' thought more directly. He tape-recorded group discussions among 3-4 pupils without a teacher being present, and then examined the number and quality of utterances in the transcripts. One of the categories he counted consisted of utterances where reference was made to out-of-school experience, either bringing it into comparison with what was being discussed in class, or interpreting it with the help of the vocabulary of the biological topic. Potentially such utterances show the learner in the act of converting 'school knowledge' into 'action knowledge'. So how common are they when the pupils are left to themselves?

PERCENTAGES OF DIFFERENT TYPES OF UTTERANCE USED IN SMALL-GROUP DISCUSSIONS OF BIOLOGICAL TOPICS.

	GROUP I	GROUP II	GROUP III	GROUP IV
Percentage of utterances containing no reference to out-of-school experience	78 %	70 %	82 %	92 %
Percentage of utterances containing reference to out-of-school experience	22 %	30 %	18 %	8 %
Total number of all utterances per minute.	18	23	14	13

Age of pupils - 14+ Size of groups: I - 4 pupils III - 4 pupils
II - 3 " IV - 3 "

Topics discussed:

I - the characteristics of xerophytes in comparison with mesophytes

II - the features of a selection of organisms which may be used for their classification

III - how do stomata open and close?

IV - the possible donations of human blood from donor to recipient.

Typically, there are 17 utterances per minute.

This work certainly indicates that certain kinds of classroom activity can involve the utilization of outside knowledge for consideration alongside the school topic. It offers a means of exploring the conditions under which it could happen. A research programme as well as a teaching policy could be developed from it. Is the reciprocal implication valid, that school knowledge will be imported into the "discussion" in later life? Does that need to be tested directly by interviews, questionnaires, and other suitable techniques?

M. Hornsey[15] has tape-recorded speech in science classes, where the teacher goes round from group to group while practical work proceeds. He prepared complete transcripts of science lessons (22 hours, six different teachers), examining all the pupil speech that occurred in direct interaction with the teacher. (Contrast Horsfield's method where the teacher is deliberately absent.) Under these conditions he found that approximately 10 % of the pupils' utterances involved experience outside the class, anecdote, analogy or comment not based directly on the teacher's contribution. In most cases the teacher acted as a sympathetic audience and then used the opportunity to add information. (Consider the alternative role of getting the pupil to go on talking, to clarify further the incipient connections in his thought.)

Areas of Opportunity

Peter Kelly suggests that 'biological concepts have rarely featured in studies of cognitive development, and that there is a need for more studies of how the form of teaching varies with the concepts being taught and how this influences the selection and understanding of the concepts by the pupils.'

He takes up Pringle's idea that there is a 'dichotomy between those aspects of biology which are similar to those of the physical sciences and those which clearly are not.' He suggests that there are 'two categories of biological concepts - biophysical and biosocial.' The former relates to physical science-based biology, and the latter to studies of social and evolutionary perspectives of life.' He ends his paper by suggesting that different styles of teaching are necessary for the different categories of concept.

It that were so, it is important to identify and distinguish the different types of concept. Though we don't fully agree with the categorisation, we have nevertheless found the distinction of biophysical and biosocial concepts helpful in provoking thought.

In the light of this distinction, we listed some of the concepts in the Nuffield O-Level Biology course (revised version) and tried to identify some of the ones we thought socially important: that is, concepts that could enlighten opinions and decisions that people may hold or take after they have finished school. Further, we attempted an initial with life outside school and those usually seen as being concerned with school knowledge only.

Table I
SOME SOCIALLY IMPORTANT CONCEPTS FROM NUFFIELD O-LEVEL BIOLOGY COURSE (Longman Group Limited, 1975)

Text 1	Text 2
sex	nutrition
menstrual cycle	complete diet
fertilization	water balance
embryo	fluoridation
growth	dental decay
microbes	photosynthesis
pasteurization	blood groups
Text 3	**Text 4**
ecosystem	variation
habitat	immunization
succession	immunity
interdependence	inheritance
biological control	probability
community	environment
trophic levels	adaptation
energy flow	selection
food webs	antibiotic
pollution	gene
feedback	enzyme
instinct	development
social behaviour	interaction
	evolution

Those underlined, though having important links with life beyond school, are sometimes considered as school knowledge only.

Each of the concepts in Table I may be examined more closely, considering in turn the out-of-school experience that pupils could bring to bear on it as they learn, and the links with later life that the concept could have, i.e. its usefulness, relevance and applications. For instance, four of these concepts are examined in this way in Table 2.

Table 2.

CONCEPT	Out-of-school experience that could be brought to bear on it as it is learned. (Examples only *)	Areas of citizenship where the concept my have influence.
Microbes	Food that has gone "off" at home. Holiday experiences of upset stomachs. Newspaper reports on food poisoning. Flu epidemics. People having measles at the same time. etc. etc.	Food hygiene. Personal hygiene. Ensuring that precautions are taken over holiday insurance and health hazard immunization of their children. Care with infectious disease. Lack of fear of non-infectious afflictions.
Dental decay	That some pupils have more holes in their teeth than others. That some have more fillings or gaps in their mouths than others. Experience at the dentists. Sweet eating habits. Teeth cleaning habits. etc.	Throughout their own life in dental hygiene. Their behaviour regarding their own children's sweet eating habits, teeth cleaning habits. Regular dental check-ups. Attitude changes to dentist etc.
Pollution	Local knowledge of pollution. Knowledge where action has resulted after pressure groups have been formed. Industrial pollution. Pollution of air and water. Noise pollution. etc.	Making them more aware of the pollution in the local community, nationally and globally. Affecting their behaviour on issues such as nuclear power stations, nuclear waste, airports and other noise pollution.
Inheritance	That they resemble their parents/grandparents in some ways. That some traits are like one parent, others like the other. Country children are likely to have some idea of dominant genes in cattle, e.g. Hereford x Friesan - white faced, white eyelashes, black body, better for beef than milk.	Better understanding of role of inheritance and environment. Realization that certain traits are inherited, and that others are not. Heightening social awareness about whether one should have children or not.

* Clearly an important part of the argument developed in this paper is that the out-of-school experience brought by particular children is to some extent unpredictable.

Classroom Techniques

Given the theoretical position developed earlier in this paper, how should the form of the teaching be modified in respect of the concepts listed? The teacher has at least two jobs to do:

- On the one hand he has to help the pupils to a greater precision in the use of the concepts (their public versions). How does infection differ from injection? What exactly is immunization? What is sterilization (of food)? What does pollution mean? To what extent is an out-of-school statement like "They put chlorine in the water to purify it" acceptable or not acceptable?

- On the other hand he has to maintain their awareness of the other associations they have to the concept, and their willingness to explore the relevance of the refined concept in each of the situations that seem linked to it. In this aspect he is concerned with developing the richness of the meaning of the idea to each individual.

Classroom practice which takes account of these aims is likely to involve a diversification of teaching methods, to include:
A) some of the teaching techniques already in use,
B) some others - for example:
i) More free-ranging discussion, either with or without the teacher. (But how much time should be devoted to it, and what are the best ways of ensuring that it is productive?)
ii) More problem-solving tasks that use situations from life.
iii) More reflective writing in which the pupil is asked to re-order what he knows from school in relation to what he meets beyond the school.

The key to connection-making is probably active language use by the pupil. In discussion, in writing in his own way, in interrogation of the text that he reads, in reformulation of what he knows, and statement of what he wishes to know, over and over again. Expertise amongst teachers in the management of such active language tasks is improving. For example, the Teacher Education Project[16] has now published trial versions of booklets on how to organize classroom discussion and how to set written work which encourages connection-making. There remains the large problem of helping more teachers to acquire the skills of managing this sort of work, and of their gaining the freedom to do it in school.

However, if these skills are acquired, the teacher's role would develop to include one of diagnosing existing conceptual patterns and provoking and monitoring their successive modifications.

Conclusions

Reference was made earlier to the mental spectacles through which people view the world. How they see things determines the strategies and tactics they use to solve problems and make decisions, and influences the opinions they hold. Whether we like or not, teachers are in part responsible for how their pupils do these things.

We don't of course want to develop young people who view things in a fixed way - with just one cognitive structure - nor people who have no framework, no understanding of the relevant public concepts, but just a whirring pool of ideas and beliefs. The ideal is presumably people who can consider events, problems, etc., from a variety of viewpoints, and who are therefore practised in bringing the concepts of the school subjects to bear on particular problems. Logically it seems likely that they will have to practise that skill in the school. Theoretically, the act of doing so could produce an inter-connectedness of thought which would assist retention as well as use of the ideas taught in school. It might also produce flexibility in thought.

To change the metaphor, consider knowledge as a sort of sol, capable of 'gelling' to produce a structure of thought relevant to a particular life-problem. Where people consider this linking, or that pattern, there is a pulling out of a framework (the gel), which is examined critically for its usefulness in solving a particular problem. If found wanting, it is replaced, or just part of it is put back in the sol-state pool of knowledge and experience and an alternative withdrawn for further consideration.

Teaching for action beyond school seems an exciting possibility, if that is the kind of result we have in mind. However, little real progress in making this the rationale of science teaching is likely to occur until we develop:

i) adequate diagnostic techniques, and

ii) expertise in managing active language tasks which provoke connection-making in the pupils' understanding.

References and Notes

1 SCHAEFER, G. (1979), "Concept Formation in Biology", European Journal of Science Education, Vol. I, Nol. I, pp. 87-101.

2 KELLY, P.J. (1979), "Teaching styles and cognitive selection related to different types of biological concepts". Paper presented at the Cognitive Research Seminar on Science and Maths Education, Leeds University.

3 A technique which such a teacher might try is that of getting pupils to list differences between the various meanings of words. What are the key differences between growth of tissues and growth of cities, or between evolution of mammals and 'evolution' of motor cars?

4 KELLY, P.J. (1977), Science, Life and Education, Chelsea College, University of London, Inaugural Lectura.

5 SCHAEFER, G. (1979). "Concept Formation in School between Scientific and Community Demands". Paper presented at the Cognitive Research Seminar on Science and Maths Education, Leeds University.

6 This quotation is from William Whewell, writing in the early part of the nineteenth century. See Sutton, C.R. (1980), "Science,

Language and Meaning", School Science Review, forthcoming. This article traces the origins of the movement for strict definitions and 'objective' language from the mid-seventeenth century onwards.

7 BARNES, D. (1976). From Communication to Curriculum, Penguin, pp. 79-85.

8 DONALDSON, M.M. (1978), Children's Minds, Fontana.

9 BERNSTEIN, B. (1971), "On the classification and framing of educational knowledge", in Michael F.D. Young (ed): Knowledge and Control, published for the Open University by Collier-Macmillan.

10 NOVAK, J.D. (1976), "Understanding the Learning Process and Effectiveness of Teaching Methods in the Classroom, Laboratory, and Field", Science Education, 60(4), pp. 493-512.

11 SUTTON, C.R. (1980), "The Learner's Prior Knowledge", European Journal of Science Education, forthcoming.

12 WEST, L.H.T., Monash University, Melbourne. Private communication.

13 FISHER, R.A. (1979), An Investigation into Methods of Probing Cognitive Structure, MEd Studies dissertation.

14 HORSFIELD, J.C. (1979), Pupil Control Groups: An analysis of small group discussion work in biology, MEd Studies dissertation.

15 HORNSEY, M., Homerton College, Cambridge. Private communication.

16 Teacher Education Project Language Study booklets, available from the University of Leicester School of Education:
Setting Written Work
Responding to Written Work
Talking to Learn
Understanding and Leading a Discussion
Would you read this?
Redesigning a Lesson.

See also:

TOMLEY, D., "The Content of the Curriculum: The Case of Science" in The Curriculum: Lessons of a Decade, (ed) M.J. Galton, Leicester University Press, forthcoming.

EDUCATION AGAINST RACE PREJUDICES AS A TOPIC OF BIOLOGY EDUCATION

by U. Kattmann [1)]

> It was not science and knowledge of races, but ignorance that caused racism to develop.
> (E.v.Eickstedt, German anthropologist, 1949)

If the sentence quoted above still holds true, instruction and information about different races are among the most effective weapons against racial prejudices.

1. "Races" as a Subject in Biology Education

1.1. Biological classification of races and race problems. The classification of humans into races is part of the biological concept of Man..

Race prejudices and social discrimination do not arise for biological reasons; they are meant to assert the prejudices and interests of one's own group against a "foreign group".

So the reasons for race problems are to be found in economic and social antagonisms and in the power of one group over another. Discrimination of the same kind can therefore quite as well be the result of other than racial differences, religious ones, for example. Nevertheless, biology cannot remain indifferent towards racial problems, for it is mostly with biological arguments that racial discrimination is supported.

The critical point seems to be the emotional link between racial characteristics and a valuation which is rooted in social conflicts between racial groups. As a result of this connection, racial characteristics can then become the strongest discriminatory factor. People with dark skin, for example, are regarded as inferior, are given worse accomodation and are cut off from the better paid jobs. Together with the colour of the skin, the colour and form of the hair can determine the place of a person on the social scale. Biological differences are emphasized and abused for giving social conditions the pretence of being natural.

Biologists may not be able to prevent each abuse of biological knowledge; the example of race problems shows, however, that there must be something wrong with biology itself, if it always emphasizes the differences and inequality between men and seems to offer more arguments for discrimination than against it. For in doing so, biology neglects the variability of races.

---

1) Paper first presented at the 7th AABE "Biology Federal and Community Development", Kuala Lumpur 1978.

1.2. The biological concept of race. In older concepts and definitions, race was regarded as a collective name for individuals whose characteristics were as much alike as possible. This concept was best represented in the races of domestic animals like dogs or horses, which are bred to achieve certain characteristics and standards and to which the term "pure race" applies. For geographic races, such as those of man, these typological terms are completely out of place.

The modern concept of race is one of statistics and population genetics. Populations are groups of individuals belonging to a species, whose members can reproduce among one another, while they are more or less separated from other populations by geographic (and, in the case of man, cultural) barriers. A division of a species into races has to start from these populations and not from individuals.

Races can be defined as populations of one species differing in their gene frequencies. So the members of a race are not all alike, but differ from one another. Quite often the variability within a race is greater than the (average) differences between the races.

The concept of variability opens up a fundamentally new way of judging and valuing racial differences. Therefore biology education can out of its own disciplinary approach contribute convincingly to an education against prejudices by a sufficiently early and pointed encouragement of population thinking. This does not see biological differences as a deviation from one's own norm, i.e. the norm of one's own group, but regards variability as a condition for the continuation and evolution of all mankind. At this point, the educational aim of instruction is directly connected with the scientific interpretation of racial differentiation.

1.3. Social equality and genetic differences. Equality of the races is a social concept concerning the individual and does not at all mean genetic equality of groups. The idea of the quality of all mankind is so very important and significant because all human beings are different from one another. If they were all alike, the concept would be superfluos and meaningless. So equality is not the same as identity, and the genetic differences between people do not lead to an inequality of rights or claims. Chances should be equal for everybody, so that all can develop differently, no matter which race they belong to. This is no contradiction, for 'equality of all human beings' does not mean making the individuals all the same but giving them all the same chance to develop in their own way. According to this, genetic inequality calls for equality of chances and equality of rights and should be a reason and an inducement to fight against restrictions or prejudice directed towards any group.

Prejudice and discrimination against foreign races in our societies cannot be justified by means of biological arguments of the natural order of things.

1.4. Which is the right age for education against prejudices? According to the latest socio-psychological investigations, no inherited defence reactions against people of other races seem to exist. The child, in the course of his socialisation, learns to accept the norms of his own group and thus assumes the values and prejudices of his environment. Rather than particular modes of behaviour, a general capacity to learn is inherited. Some behavioural scientists assume that Man possesses learning dispositions with which he identifies himself emotionally to his own group and rejects strange groups. The assumption of genetically determined learning dispositions merely means that learning which corresponds to the dispositions is particularly favoured, but is by no means dependent on them if different learning conditions are brought into play. Group thinking, moreover, leads to a more or less positive assessment and description of one's own group (self-image, auto-stereotype) and a more or less negative assessment or description of strange groups (stranger image, hetero-stereotype). This tendency is easily recognizable in, among other things, the increasing orientation towards one's own group (cf. Goodman 1964, Clark and Clark 1947, Morland 1962, Vaughan 1963; see also Harding et al. 1969, 18-21.

Investigations indicate a distinct group and class specify of prejudices against people of other races. Some studies find this already present in children of pre-school age, and all studies confirm the increase in negative prejudices between 8 and 12 years of age. Then, between the ages of 11 and 16, this becomes as distinct as in adults (Allport 1954, Blake & Dennis 1943, Vaughan & Thompson 1961, see also Harding et al. 1969, 17-24, Kattmann 1977).

According to these results, instruction which intends influencing the attitudes of pupils towards people of different races should begin early.

On pre-school and primary school level, when the child starts orienting himself toward his own ethnic group, we should try to find a way for him to identify himself with his group without developing animosity or contempt for foreign groups (concept of equal value of all groups of man).

Especially in the grades 4 and 5, when the child develops distinct stereotypes, we should start on breaking those down or keeping them from becoming rigid. In this we are helped by the ability to differentiate which is developing at this age. At this age, differences between people can be emphasized by instruction and regarded as positive (concept of the variability of man).

In order to train the students in population thinking, it is also necessary to repeat the topic of variability on different levels.

The guiding question for the relevant biology lessons could be: "What is the significance of man's variability?" (see table). This question is part of the structuring principle of the human-centred approach for biology education; (cf. Kattmann and Schaefer 1978). The concepts derived from this question put the subject of races into a wider context, where the biological question of the statistical distribution of characteristics is pursued on various levels (evolution, population, social groups, sexes, individuals) up to the domain of the social sciences.

What is the significance of man's variability?

Concepts:

1. Future evolution of mankind (family planning, optimalisation of social conditions, including: diseases of civilization and by inheritance, eugenic family consultation, environment planning).
2. Biological differences between races (racial differences and variability within races; concept of race according to population genetics; psychological differences between races, race and cultural development).
3. Biological differences between social groups (social types, screening, selection by social conditions).
4. Biological differences between the sexes (sex differences and their variability; sex roles and their dependence on socio-cultural factors).
5. Biological differences between individuals (variability in physical and psycho-intellectual characteristics).

2. Effects of Education Against Race Prejudices as Shown by the Biology Teaching Unit "Man and Animals"

2.1. The teaching unit "Man and Animals". Traditional biology education in Europe often gave the impression that "Man" meant "white man" or "European". This notion came about unintentionally as a result of Man being considered only from morphological and anatomical 1 points of view, and because only "white" men were shown in illustrations.

In order to replace the unicoloured image of mankind by a multi-coloured one, a team at the IPN in Kiel developed the instruction unit "Man and Animals" (Kattmann and Stange-Stich 1974 , see also Schaefer and Kattmann 1976). The unit is concerned with conveying the concept of Man to ten to eleven year-old pupils in a manner suited for this age group.

In the first part, the biological species "Man" is considered in comparison to animals so that the single species including all variations is contrasted with the diversity of animals. The second part deals with the division of the species of Man into races. It consists of four lessons covering the following subjects:

(1) Mankind consists of many races.
(2) The division of mankind into large races.
(3) Racial problems.
(4) What the races have in common and how they can live together.

In spite of the unequivocal objectives of replacing racial prejudice by rational judgement, some teachers believe that biology education on races may actually arouse new or reinforce old prejudices. It was, therefore, most important to prove the effects of teaching.

2.2. Method of investigation. The investigation was to answer two questions:
(1) What prejudices do the pupils in our group have in 5th grade?
(2) What effect does teaching about races have on these prejudices (cf. Kattmann 1974)?

In order to be able to answer these questions, we selected a simple and often used procedure: the semantic differential. The procedure seemed suitable to us for use in questioning, since it is difficult for the pupils to regard it as a poll of their opinions. For this reason, in comparison with other procedures such as questionnaires and interviews, the answers are probably less tailored to fit the teacher's expectations. The special semantic differential and the layout of the test sheets were developed by Stange-Stich according to Hofstätter (1953) and Osgood, Suci and Tannenbaum(1957). The sequence of the test sheets was systematically varied.

The task of filling in the semantic differential sheet consists in judging certain concepts emotionally on the basis of a series of antonyms (polarities). "Profiles" of the tested concepts are obtained. Their similarity to one another can be expressed as a correlation. These correlations express the "affective similarity" which the concepts possess in the pupils' judgement. In our sample, the concepts "Man", "Chinese", "Negro", "German", "tenderness", "joy", "dirt", "cruelty" were tested.

The samples were carried out on two groups of 5th grade pupils. Group E comprised six classes of the integrated comprehensive school at Neumünster, and about 170 pupils. Group C comprised two classes each from a secondary modern, a secondary technical and a grammar school in Kiel, altogether about 150 pupils.

2.3. What prejudices exist in 10 and 11 year-old pupils against people of other races?

The point of departure for the investigation was that the pupils do not often classify people of other races under the concept "Man" and that they reflect the negative social prejudices against people of other races. An opinion about the concepts "Negro", "Chinese" or "German" which deviates from the concept "Man" is regarded as a sign of the existence of a prejudice regardless of whether the opinion is a positive or a negative one.

The direction of the valuation can be seen from the polarities themselves. It can also be seen from the differing correlations of each particular race to the positive concepts "tenderness" and "joy" and the negative concepts "dirt" and "cruelty".

The values of the total correlations between the tested concepts show that among the concepts which refer to

"Negro", "Chinese", "German" the concept "Negro" is valued most negatively. The concept "Negro" is correlated less often to "Man" and to the positive concepts ("tenderness", "joy") than "German" and "Chinese" are. There are no negative correlations between "Negro" and the negative concepts ("dirt", "cruelty").

The primary hypothesis, from which the presence of negative prejudices against people of other races was conjectured, is thus reinforced in the case of the concept "Negro", but less so in the case of the concept "Chinese".

The different opinions about "Negro", "Chinese" and "German" under the concept "Man" are clearly shown by a comparison of the mean values of the 13 polarities in which the concepts "Man", "Chinese" or "Negro" are significantly differentiated (see Diagram 1).

"Chinese" is nearer to the concept "Man" than "Negro" in all polarities (cf. particularly the polarities: clean - dirty, savage - gentle, kind - cruel, peaceful - hostile, foolish - wise).

The results of the tests indicate the existence of prejudices of a negative stereotype in the 10 to 11 year-old pupils of our populations.

2.4. What effects does instruction have on pupils' prejudices?

In groups E and C the test was repeated after about three months. Group E was the experimental group to which the unit "Man and the Animals" had been taught in the interval between the two tests. Group C was the control group to which the normal course, without any reference to race, had been taught.

The design of the investigation shows that this was not an experiment in which particular variables in the experimental group were varied. It was more like an accompanying investigation. The test was meant to show whether instruction has positive or negative effects on the pupils' attitudes. The control group served mainly to illustrate what alterations were caused solely by the repetition of the test. At the same time, those changes due to factors other than school, i.e. T.V., newspapers, could be covered.

As far as the experimental group is concerned, changes essentially concern the concept "Negro". These changes are not observable at all in the control group.

According to the results of the first tests, the pupils of both groups had negative prejudices against "Negro". The aim of instruction was at least to reduce these prejudices. The second test with the experimental group was expected to show such a change of attitudes in a closer correspondence of the values ascribed to "Negro" and to "Man". This is clearly the case with the correlations of "Negro" to "Man", as well to "German", "Chinese" and the positive emotional concepts "tenderness" and "joy". The changes in the negative emotional concepts also agree with this tendency, even though these values do not differ significantly from those of the control group.

This result is also confirmed by the average values of the polarities for the concept "Negro". Eight of these values differ significantly in the second testing from those of the first one. Each and every one of these alterations in the average values indicates that their new level more closely approximates those of "Man" (see Diagram 2, cf. the polarities: savage - gentle, kind - cruel, peaceful - hostile). None of the "Negro" profiles in the first and second samplings of the control group exhibit any significant alteration in the average polarity values. The results, therefore, generally confirm the assumption that the pupils' attitudes toward "Negro" were improved by instruction as proposed in the unit "Man and Animals". The counter-assumption that prejudices would be reinforced did not hold true.

A further result in the affective domain is the essential assimilation of the opinions about the two groups of strangers ("Negro", "Chinese") brought about by instruction. "Negro" and "Chinese" correlated almost equally with "Man" after the lessons.

Instructions hardly had an effect on the almost identical valuation of "German" to "Man". One suspects that the pupils continue to feel that both these concepts are the same even after the lessons. The contrast between self-image (auto-stereotype) and stranger image (hetero-stereotype) was neither abolished nor weakened by this kind of instruction.

It came out that four lessons are not sufficient to achieve such a goal.

2.5. Conclusions

It has been argued that the influence of school instruction on racial attitudes is negligible and that direct acquaintance with people of other races is more important. Empirical investigations record, however, that acquaintance and contact do not necessarily reduce the level of prejudice. Negative attitudes are often reinforced as much as positive ones when meeting other people (cf. Karsten 1965, Harding et al. 1969).

Some teachers are afraid that instruction on races might provoke and strengthen rather than reduce prejudices. According to an inquiry by Trubowitz 1969, an attempt at reducing prejudices in New York pupils through outings and discussions of racially mixed groups was a failure. Miller 1969 even found significantly increased prejudices in 15 to 20 year-old students after a course of instruction which had actually been supposed to reduce prejudices.

Verma & Mac Donald (1971) and Verma & Bagley (1973) do report slightly positive effects of their lessons involving 14 to 16 year-old British students, but the results of their inquiries seemed so unconvincing to their critics that the Humanities Project (School's Council) materials on "Race" remained unpublished.

In contrast, our inquiry agrees with that of Glock et al. (1975), according to which information and c ontroversial discussion operating together can markedly reduce prejudices. Both these methods of teaching were used at crucial points in the instruction unit "Man and

Animals". The study demonstrates that, contrary to popular belief and educational practice, group and race differences ought to be acknowledged and discussed - even when the information is expected to be negative. Contrary experiences of other educational projects are not inconsistent with this: in the future we shall have to find out which aspects of instruction can change prejudices in such a way as to comply with the relevant learning objectives. One point of interest will be to determine which age is the right one for each of the different educational subjects and methods.

The results of our study indicate that it is easier to achieve a change of attitude in younger pupils (11 years of age in our case) than in older ones (14 years of age in the inquiries of Miller 1969, Verma and Mac Donald 1971, Verma and Bagley 1973). This confirms the findings of inquieies on the development of ethnic stereotypes (cf. Kattmann 1977).

Prejudice is reversible. Biology education can help to reduce prejudices if problems and differences are not passed over in silence, but discussed freely so that information and discussion can pave the way to understanding and a critical review of attitudes.

References

ALLPORT, G.W. The nature of prejudice. New York: Addison-Weseley, 1954.

BLAKE, R., DENNIS, W. The development of stereotypes concerning the Negro. Journal for Abnormal and Social Psychology, 1943, 38, 525-531.

CLARK, K.B., CLARK, M.P. Racial identification and preference in Negro children. In T.M. NEWCOMB, E.L. HARTLEY (eds) Readings in Social Psychology. New York: Holt, 1947, 169-178.

EICKSTEDT, E. VON Vom Wesen der Anthropologie. Homo, 1949, 1, 1-13.

GLOCK, C.Y., WUTHNOW, R., PILLIAVIN, J.A., SPENCER, M., Adolescent prejudice. New York: Harper & Row, 1975.

GOODMAN, M.E. Race awareness in young children. New York: Crowell-Collier, 1964.

HARDING, J., PROSHANSKY, H., KUTNER, B., CHEIN, I. Prejudice and ethnic relations. In G. LINDZEY, E. ARONSON (eds) The handbook of social psychology. Vol. 5, Reading, Mass.: Addison Wesley, 1969, 1-76.

HOFSTÄTTER, P.R. Gruppendynamik. Reinbek: Rowohlt, 1953.

HOFSTÄTTER, P.R. Einführung in die Sozialpsychologie. Stuttgart: Kröner, 1966.

KARSTEN, A. Die Anderen im Urteil der Jugend. München: Juventa, 1965.

KATTMANN, U. Die Einstellung elfjähriger Schüler zu Menschen fremder Rassen, empirische Untersuchungen und didaktische Konsequenzen. Westermanns Pädagogische Beiträge, 1974, 26, 255-264.

KATTMANN, U. Die Entwicklung von ethnischen Vorurteilen bei Schülern und Möglichkeiten ihrer unterrichtlichen Beeinflussung. Die Deutsche Schule, 1977, 69, 401-412.

KATTMANN, U., STANGE-STICH, S. DER Mensch und DIE Tiere. Unterrichtseinheit für die Orientierungsstufe. IPN-Einheitenbank Curriculum Biologie. Köln: Aulis, 1974.

KATTMANN, U., SCHAEFER, G. Restructuring School Biology - Relevance and Consequences of New Approaches for Biology Teaching and Teacher Training. In: HERNANDEZ, D.F., KELLY, P.J. (Eds): Sixth Asian Regional Conference in Biological Education.Proceedings of the Sixth Biennial Conference in Biological Education, July 28 - August 2, 1976, Bangkok, Thailand, 1978, pp. 47-59.

MILLER, H. The effectivenes of teaching techniques for reducing colour prejudice. Liberal Education, 1969, 16, 26-31.

MORLAND, J.K. Racial acceptance and preference of nursery school children in a Southern city. Merill-Palmer Quarterly, 1962, 8, 271-280.

OSGOOD, C.E., SUCI, G.J., TANNENBAUM, P.H. The measurement of meaning. Urbana: University of Illinois Press, 1957.

SCHAEFER, G., KATTMANN, U. IPN unit bank biology - a new type of biology curriculum. Journal of Biological Education, 1976, 10, 242-248.

TRUBOWITZ, G. Changing the racial attitudes of children: the effect of an activity program in New York children. New York: Praeger, 1969.

VAUGHAN, G.M. Concept formation and the development of ethnic awareness. Journal of Genetic Psychology, 1963, 103, 93-103.

VAUGHAN, G.M., THOMPSON, R.H.T. New Zealand children's attitudes towards Maoris. Journal for Abnormal and Social Psychology, 1961, 62, 701-704.

VERMA, G.K., BAGLEY, C. Changing racial attitudes in adolescents, an experimental English study. International Journal of Psychology, 1973, 8, 55-58.

VERMA, G.K., MACDONALD, B. Teaching race in schools: some effects on attitudinal and sociometric patterns of adolescents. Race, 1971, 13, 187-202.

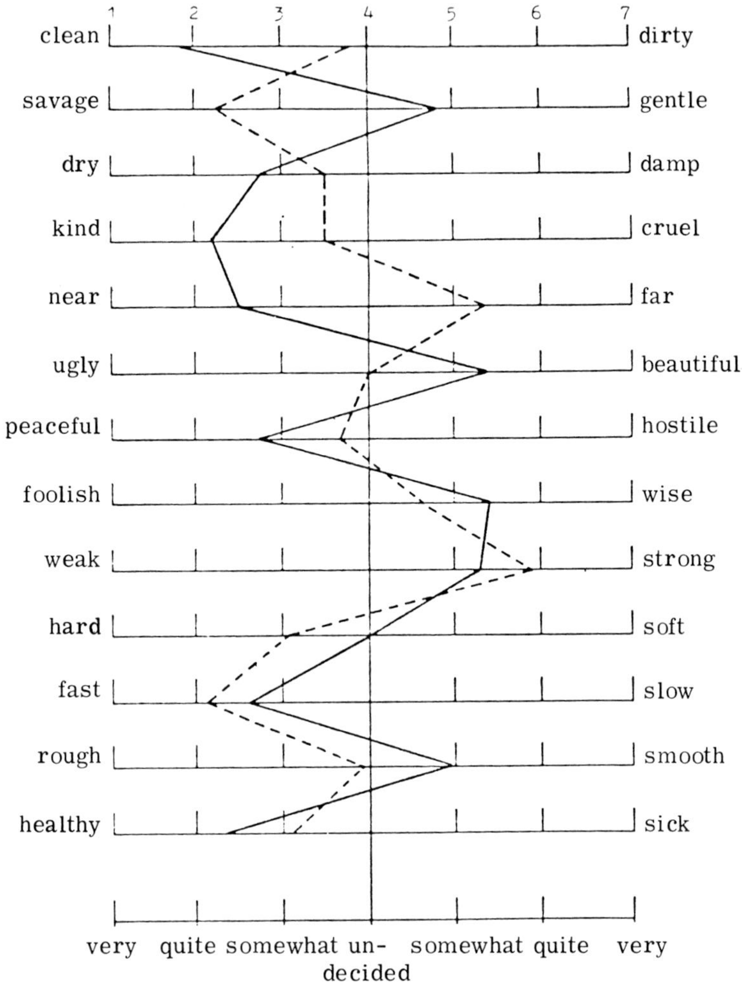

Diagram 1a): Mean values of the 13 polarities in which the concepts "Man", "Chinese" or "Negro" are significantly differentiated in the pre-test (group E):

a) "Negro" and "Man"

______________________________ "Man"

------------------------------ "Negro"

Attitude towards Races

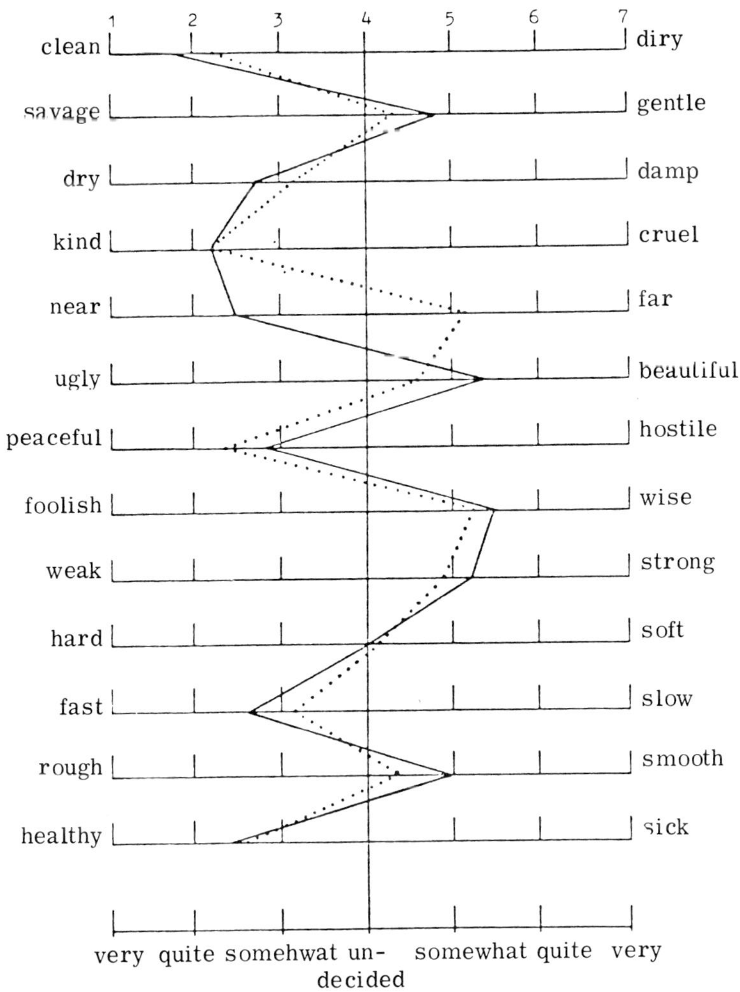

Diagram 1b): Mean values of the 13 polarities in which the concepts "Man", "Chinese" or "Negro" are significantly diffentiated in the pre-test (group E):

b) "Chinese" and "Man"

______________________________ "Man"

.... "Chinese"

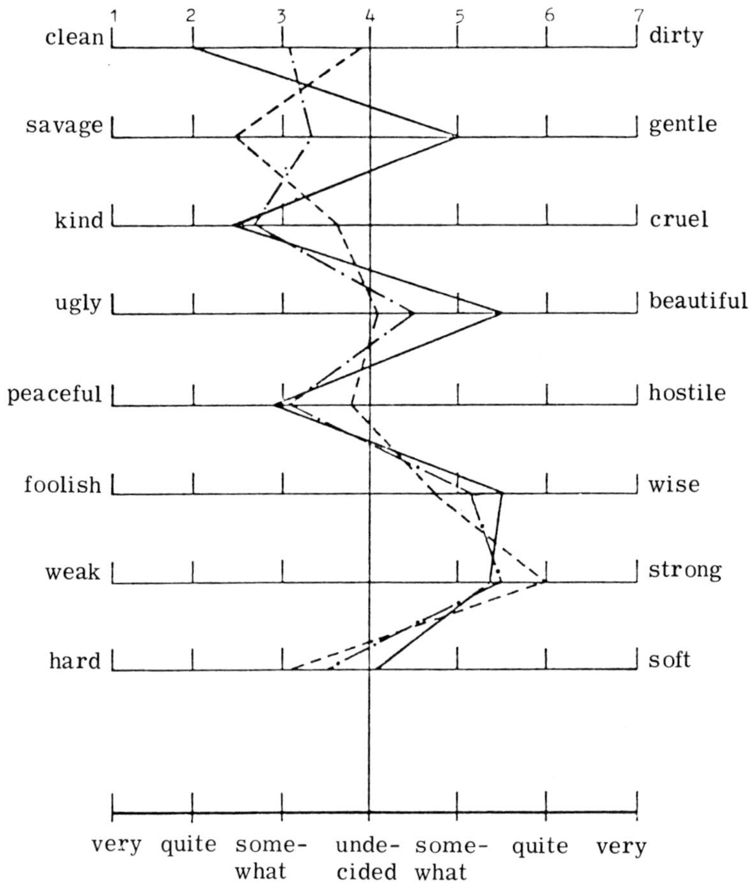

Diagram 2: Average values of the 8 polarities, whose average values differ significantly for the concept "Negro" in the experimental group (E) in the pre- and post-test. The diagram shows the average values for "Negro" in the pre- and post-test and the average values for "Man".

________________________	"Man"	
-------------------------	"Negro"	(pre-test)
-.-.-.-.-.-.-.-.-.-.-.-.-.	"Negro"	(post-test)

IMPROVEMENT OF BIORESOURCES - AN IMPORTANT ASPECT OF APPLIED BIOLOGY

by A.N. Rao

The problems of food, energy and environment are of very great importance to mankind to-day and they should be considered with utmost priority in any functionally oriented biological education. The biologists deal with the natural resources most basic to the well-being of the society. Food, clothing, housing and furniture, protection of the soils, catchment areas for the reservoirs, reservoirs for the rain water, food and cover for wild life, all these are directly or indirectly dependent on the plants as natural resources. Although considerable research has been carried out in areas such as increased food production, solar energy for photosynthesis, biogas production, recycling of agricultural wastes and others, very little of these are included in the teaching curricula of higher schools and universities of developing countries. The objective of this short paper is to highlight certain aspects of bioresources and to emphasize the need for their inclusion as part of the education in Applied Biology.

The enumeration of data and identification of bioresource materials require further work in the fields of taxonomy and other related areas. In the tropics the natural flora is extremely rich and varied and of this to-date only a small portion has been utilized by man. No doubt there is plenty of scope to select additional varieties to supplement the traditional ones already under use but also to improve the recognized varieties further. This principle is applicable to timber trees, fruit trees, vegetables, fermented food substances, fuel and foliage plants. For example of the 2000 or so tree species present in the tropical lush forests of Malaysia and Indonesia only about 100 species of timber trees are exploited as common timbers and this amounts to hardly 2 % of total tree species available. A preliminary survey conducted on the fruit trees of Malaysia, Indonesia, Philippines, and South Vietnam has shown that there are 30 genera and 131 species of fruit trees, both wild and cultivated. This would provide a suitable example to indicate the richness of the utilizable species from the flora of this area. The primitive cultivars need to be genetically improved both for yield and the quality of fruits. The scientific details of some of the wild taxa are relatively unknown to science. The different problems like geographical variation, genetics of trees, phenology, fruit set and development, conditions for their cultivation are yet to be studied. In that part of Asia where these fruit trees are common, the fruits are directly harvested from the forests and such tree species should be brought under cultivation immediately. Because due to deforestation and land reclamation, there is an imminent danger that these fruit tree species would virtually disappear from the forests before the biologist will have the chance to make use of them for cultivation or their genetic improvement (Sastrapradja, 1975, Loh, 1975). It is encouraging to note that an urgent call has been made by some of the international organizations like FAO, UNESCO and other regional institutes functioning in South East Asian Countries. An international board for plant genetic resources (IBPGR) was set up by FAO to look into the urgent problems and to conserve the genetic materials needed for future work.

About 4400 million hectares, approximately 1/3 of the total land area of the world, is covered with forest vegetation and of these nearly 28 % or 1200 million hectares are tropical forests. The annual rate of deforestation in the tropics is 15-20 million hectares per year and taking the phenomenal increase of population growth in the tropics, the forests may disappear in about 40 years or so. To prevent this catastrophic change, a well managed scientific approach is very essential based on 'multiple use' and 'sustained yield of forest products' theories.

Reforestation programmes to offset the losses and to improve the natural resources should be launched. As is well known both the climatic pattern and water resources largely depend on the existence of dense forests. Closely linked with the tropical forests is the rich variety of wild life and again some of the species are threatened as IUCN reports publicise from time to time.

Conservation and improvement of gene pool of plants to improve our biosciences deserves the highest research priority. The research work on both pure and applied aspects is important here. Plant tissue culture has helped to accelerate the breeding work of crop plants and to produce new hybrids.

The history of agriculture tells us that the modern agriculturist or plant biologist has failed to introduce even a single important new food crop. The existing staple cereals, vegetables, fiber plants, spices and beverage plants are inherited from the natives of different lands. No doubt that the modern biology has helped to improve the primitively cultivated, poorly yielding genotypes of yester years. Our continued dependence on the restricted number of genotypes of the major food plants and the improper usage of their yield may contribute to several catastrophies of great proportions. The solutions for such problems are not only the basic research on some of the tropical plants which may help us to identify new food crops but also to improve our knowledge on the post-harvest physiology of the existing crops.

Increased production of agriculture has brought prosperity for many developing countries. But the existing facilities for proper storage and transportation of the agricultural produce in the developing countries are far from adequate. This is more so with regard to the perishable food materials like fruits and vegetables. In the S.E. Asian region cold storing of local fruits and vegetables is a rare phenomenon although many fruit importers use cold storage facilities for oranges, grapes, apples that come as far as from California, Israel and middle European Countries. Even the basic information required to preserve the local fruits under refrigeration is not available. The nature of fruits and the tissues therein and how they respond to storage conditions under given temperature and humidity conditions have yet to be determined in most of the tropical fruits. The results of such studies are obvious. That is to extend the shelf life of fruits and to subject them to prolonged storage under suitable conditions so that a uniform supply of these fruits throughout the year is ensured. Clearly this is one of the areas where both the pure and applied research in the fields of botany, horticulture, and storage engineering and others should come together and the results obtained by their joint effort should help the common people. The facilities available so far are being crippled due to the new energy crisis and escalation in oil prices.

On a global scale the experts in the relevant areas have calculated that there is a 10 % loss for cereals and grain legumes and 20 % for non grain staple foods like yams, cassava, and other perishables including fish. Even these modest estimates indicate the heavy food losses on the world wide scale and these losses can only be checked by the combined work of applied biologists and industries. The UN general assembly passed a special resolution in 1975 to prevent the food losses. Theoretically, it is possible to conserve most food commodities and to prevent their loss. It is an acknowledged fact that many countries do not have the trained biologists. Instead of waiting for the local people to get trained, the help from the developed countries should be obtained. Some of the problems that deserve immediate attention may be solved by establishing co-operative research between scientists from developing and developed countries. But in many countries there are no trained biologists who can undertake the joint research work on this important problem. All the expertise brought from the advanced countries would no doubt increase the cost of operations and therefore the projects may become non-viable. A compromising solution has to be found somewhere between these two seemingly impossible suggestions. All these problems should eventually be solved by adequate biology education efforts at secondary and tertiary levels producing the required man power. In many countries, the need for properly trained technicians in applied biology is as great as the postgraduates trained in the same field and other engineering disciplines.

Applied biological research on nitrogen fixation, photosynthetic processes, salt and temperature tolerance can yield valuable data to improve the bioresource utilization further. The basic findings in the fields of genetic engineering, fermentation technology, and other related areas may bring rapid developments in the near future. Such interdisciplinary activities play a catalytic role in these efforts of bioresource development.

Thus progressive efforts are very timely and necessary to explore new bioresources and to maximise the use of the ones we already have. These aspects should also be adequately emphsized in the courses on Applied Biology offered in the various institutions of higher learning of the developing countries.

Late considerable emphasis is laid on the utilization of microorganisms in enriching the bioresources. Both the world data bank on microorganisms and the world federation of culture collections are putting forth their efforts to prevent loss and to efficiently use the stocks of certain microorganisms. Thus the gene pool conservation is a dynamic and not a static effort. The utilization of waste materials or converting the cheap food into a highly nutritional one with the use of microorganisms are the important means of enriching the bioresources. Some of the biological methods already employed by the so-called uneducated people in the rural areas and which has progressively continued for several centuries should be further improved on the basis of modern biological knowledge. It would be a great help if these methods are continued to develop as rural industries or small scale industries.

Regarding the educational aspect, hardly there is any published information about the food losses and loss reductions in the developing countries. A limited amount of data may be available for grain crops but information is particularly lacking about perishables like fruits and vegetables. Due to the good efforts of world bodies like FAO, USDA, Institute for Scientific Co-operation at Tübingen and others, some data are available but these are scattered in various technical bulletins, reports and journals, making them untenable for teaching purposes. The data available for the food loss in industrialized countries are not directly applicable to the situations in the developing countries. The situation requires cooperative work between national and international organisations to produce suitable text-books for schools and colleges. There is an urgent need to bring some of these topics for classroom biology teaching (Purseglove 1974, Frankel and Hawkes 1975).

The classical type of biological education has to be modified suitably and wherever necessary the applied aspects of biology pertaining to food, agriculture or other aspects should be adequately emphasized along with the basic biological courses. In other words, there should be a good mixture of pure and applied aspects included even for the first degree, Bachelor of Science degree biology courses. Students thus trained should be able to help the farmer, the food processor, bringing an improvement in agricultural production, food distribution and nutritional well-being. The problems of the developing countries in this respect are diverse and no definite or one pattern of curricula in applied biology may be adequate to solve all the problems of any one country, and therefore the emphasis should be varied to suit the local conditions. It may become necessary to adopt a wider spectrum of disciplines and in the planning of biological education particularly in relation to the enriching of bioresources.

In conclusion, I have attempted to remind ourselves that as biologists we can instill in our students and the public, a broad philosophy about using the natural resources in a very economic way avoiding the wastage or spoilage wherever possible. This could not only increase the value of our economic products but also enhance in many ways the lives of the present and future generations.

References

FRANKEL, O.H. and HAWKES, J.G., 1975. Crop genetic resources for to-day and tomorrow. IBP. Vol. 2. Cambridge University press. Cambridge.

LOH, C.M., 1975. Fruits in Peninsular Malaysia. 47-52. South East Asian plant genetic resources, see details under S. Sastrapradja.

PURSEGLOVE, J.N. 1974. Tropical crops. Longman, London.

SASTRAPRADJA, S. 1975. Tropical fruit genus plasm in South East Asia. 33-46. Proc. Symp. South East Asian Plant Genetic resources. Biotrop, Bogor, Indonesia. (Eds. J.T. Williams, C.H. Lamoreux and N.W. Soetjipto).

Section IV

Appendix (Additional Tables)

COMMUNITY PROBLEMS AND RELATED BIOLOGY TOPICS

by D. F. Hernandez [1)]

1. Overpopulation:
Rapid growth of population. Family planning methods and their positive effects to family growth. A growing population requires space for human activities. Control of reproduction (Sexual).

2. Housing Problem:
Overpopulation causes congestion. Poor sanitation - effect of congested community. Organisms, relation to other organisms and to their environment. Competition, survival of the fittest. Ecological balance.

3. Sanitation:
Properly constructed drainage, pit and comfort rooms reduce the spread of bacteria by their carriers. Spread of disease by bacteria can be prevented by chemical action.

4. Disease Control:
Microorganisms cause certain diseases. Microorganisms multiply rapidly under favourable conditions. Man can be an active vehicle (carrier) of disease. Precautions and preventive measures against spread of microorganisms, by proper hygienic practices, is an effective means to control disease. Vaccination and immunization reduce the outbreak of diseases.

5. Malnutrition:
Concepts of growth and development, and effects of proper nutrition. Food supply in relation to man's demand. Basic food groups and their importance to the human body. Foods are nutrients, but not all nutrients are food. Food: the main source of nutrients needed by the body. The body requires a balance of organic as well as mineral ingredients of food. Certain nutrient deficiencies cause specific disorders in the body. A malnourished child is susceptible to infection. Feeding with a well-balanced diet increases resistance to infection. Severe malnutrition during embryonic development can produce congenital abnormalities in the newborn. Energy, protein and fluid intake should be increased for infected or ill persons. Vitamins as a food supplement are necessary for growth and development of an individual.

6. Food Shortage:
Plants which are sources of food should be kept under certain optimum factors. Genetic concepts (e.g., breeding, which affects the quality and quantity of food sources). Modern technology in farming, applied in a reasonable way, is superior to the traditional one. Multiple cropping pattern increases production geometrically.

7. Food Spoilage:
Microorganisms that utilize food as source of their own nutrition cause spoilage of the food. Food is chemically broken down and decay sets in when heterotroph microorganisms respire. Food

---

[1)] Prepared by 10 biology teachers enrolled in Masteral program, University of the Philippines

preservation - pasteurization, dehydration - kills harmful micro-organisms resp. inhibits their growth.

8. Grain Deterioration:
Improper drying may cause seed germination. Moist conditions are favourable to fungi attack. Infestation by insects - biological hazards for grains.

9. Pollution:
Sources of pollutants. Biological effects of radiant energy. Biological effects of exposure to air pollutants. Effects of pollution to man and other living organisms. Organic and inorganic pollutants. Pollution's adverse effects threaten the quality of man's environment. Advanced technologies are vital factors in pollution. Need for a balance between living organisms in the environment (biological equilibrium). Pollution - a consequence of overpopulation. Pollution deteriorates man's and other organisms' ecosystem.

10. Agricultural Practices:
The traditional practice of growing similar crop plants season to season and year to year depletes the fertility of the soil and hence reduces crop yield. Crop rotation with legumes (especially) restores soil fertility. Application of inorganic or organic fertilizers is a means to increase the viability of the soil.

11. Pest Control: Identification of harmful and useful insects. Life cycle of insects. Effective means of control.

12. Erosion:
Men are responsible for much erosion by overgrazing, kaingin system and destruction of forest cover. Contour ploughing and terracing slows down runoff and reduce erosion. Strip cropping impedes drainage down the slopes and conserves the soil. Reforestation with selected plant species reduces erosion.

13. Illegal Fishing:
Uncontrolled or improper practices diminish future fish harvests. Life cycle of fishes. Interdependence of fishes with their environment.

14. Problems Related to the Ecosystem: forest denudation, floods, overcollection of plants and animals, kaingin (slash and burn farming). Forest denudation/deforestation and preservation. Ecological balance and succession. Flood control attributed to forest deforestation - interaction of biotic and abiotic factors. Extinction of some plants and animals. Preservation of species and its habitat. Ecosystem as the biological capital - where total productivity depends on. Water - plant relation (forest/flood interaction). Forests as sources of food, shelter and clothing for animals. Forests as sources of water for rivers and streams. Tree planting helps restore ecological balance. Burning of grasslands (cogonal) increases the population of cogon in the area. Planting of shrubs or trees (productive or protective or combination of both) reduces the growth and population of cogon in the area. Interplanting the trees with crop trees like cassava, calamansi and others increases the marginal productivity of the ecosystem.

Overgrazing produces bare patches of ground producing gulleys by rainfall run-off. Dumping of pollutants in the water are detrimental both to the primary and secondary producers. Reduction in the viability and productivity of rivers and streams are due to unwise dumping of materials and "over-fishing".

15. Social Problems:

Alcoholism and Drug Addiction:

Biological explanations of behaviour.

Crimes:

Interpreting and predicting behaviour through genetics.

Psychological and Sociogenetic Problems:

Biological Engineering. Heredity (behaviour inheritance). Understanding the biological roots of behaviour.

ANALYSIS OF DATA TO IDENTIFY CURRICULUM NEEDS/NONFORMAL EDUCATION NEEDS
SAMPLE COMMUNITY: FISHING VILLAGE [1)]
by D.F. Hernandez

Data	Problems	Implications for Curriculum	Implications for Non-formal Education (adults)
A. Analysis of drinking water from all sources revealed infinite number of coliform bacteria per 100 ml of water.	Source of drinking water: shallow wells and a few tanks for catching rain water running off from roofs of some houses with galvanized iron roofs.	Awareness of microorganisms in the community's water supply. Knowledge about micro-organisms and spread of disease through water-borne organisms. Methods of keeping water safe to drink. Solar distillation of water to demonstrate pure water, water cycle and as source of pure water for emergencies.	Contact National Irrigation Administration to present community's problem. Demonstration of ways of making water potable: boiling, using clean containers. Sanitation activities to keep water from fecal contamination. Results of water examination discussed with adults; implications to health also discussed.
B. % of respondents wanting skills and knowledge in: 78% - fishing gear, methods. 61% - controlling respiratory water-borne diseases. 56% - controlling pests and other insects 56% - good water supply	Low catch of fish. Small size of fish caught. Idle unrepaired machines. Children frequently ill. Large populations of insects: mosquitoes and other pests.	Comprehension of biology of local fishes (life cycle, habitat). Identification of common fishes of the locality. Methods of fish capture. Awareness of need for preservation of fishing grounds, avoidance of overfishing. Knowledge of means of preserving fish applicable to locality. Knowledge about parasitic diseases.	Seminar workshop on fishery education using local resource persons from government fish extension bureau/projects in nearest city or town. Content of seminar included: improved fishing methods and gear, e.g. making encircling gill nets, outfitting fishing boat with pump engine for deep-sea fishing. Clean-up campaigns of entire community, especially garbage dumps. Seminar on health and nutrition.

1) Based on a project actually conducted by the Science Education Center.

Data	Problems	Implications for Curriculum	Implications for Non-formal Education (adults)
C. Use of fine mesh nets by 26% of the respondents.	Illegal fishing.	Knowledge of proper utilization of natural resources (conservation). Knowledge of the ill-effects of illegal methods of fishing.	Demonstration of fish preservation. Introduction of fish farming (putting up a fish pen). Maintenance of fish pen.
D. Twenty-four toilets in a community of 232 families.	Lack of sanitary waste disposal.	Awareness of need to dispose human wastes in a sanitary way. Knowledge about micro-organisms/parasites causing infectious diseases and parasitism in the population.	Seminar and demonstration on construction and use of water-sealed toilets by R & D team. Encourage use of services from Rural Health Unit personnel (local) (latest count revealed a total of 165 toilets).
E. There is no doctor in the community.	Inaccessibility of medical service/health care.	Awareness of need to select the right people for health services. Knowledge of the use of medicinal plants for simple ailments. Need to maintain a medicinal plant garden in school. Awareness of difficulty of getting proper health care: need to keep oneself healthy.	Seminar on health education and proper care of sick children (with help of local consultants from Ministry of Health and Ministry of Education and Culture).
F. Soil analysis showed only a few kinds of plants can grow.	Limited vegetable supply and hence no or limited vegetable in the diet resulting to unbalanced meals and poor nutrition.	Awareness of need for balanced meals for proper nutrition. Knowledge of the importance of vegetables in the diet and good nutrition. Awareness of the kinds of plants viable in the area: alugbati, beans, squash, kadyus & other legumes, etc.	Seminar with mothers on preparation of nutritious foods locally available. Community campaign on vegetable gardening in trench and used containers (plastic, bamboo rafts or pots). Plant malunggay project (a fast growing tree, the leaves of which serve as vegetable)

Data	Problems	Implications for Curriculum	Implications for Non-formal Education (adults)
G. Interview showed that 72% of respondents want to learn how to repair & maintain pumpboat engine.	Repair & maintenance of pumpboat engine.	Knowledge of the parts and uses of a pumpboat engine (in practical arts or vocational class).	R & D contacted experts in engine repair & maintenance. Conducted seminar & practicum on the topic. Additional or improved skill of fishermen on engine repair & maintenance, hence no need of hiring other men to do it for them (self-sufficiencies).
H. Literacy rate is 95%; adults want to read on fishery, health & other topics pertinent to their needs.	No accessible literature or reading center. Lack of proper type of informative materials.	Awareness of improving literacy by reading. Supplement knowledge learned in school by reading. Perusal of nonformal education materials will make children aware of community concerns.	Providing reading materials on various topics for children & adults to put up reading center - as a reinforcement of functional literacy and to increase knowledge on topics relevant to their needs. Contact extension units of government, private agencies which may continue providing reading materials.
I. No leisure time activity except visiting neighbours, playing cards.	Lack of recreational facilities/activities.	Awareness of the need for wise use of leisure time. Knowledge of wholesome leisure/recreational activities. Physical fitness: sports. Reading for pleasure.	More productive use of leisure, i.e. mending nets, cottage industry (fish preservation for market). Physical fitness: sports. More reading centers set up with wider variety.
J. Survey showed that this is a sustenance fishing community.	Lack of capital.	Knowledge about cooperatives, how these work, their advantages. Knowledge about making bank loans.	Contact Ministry of local Government & Community Development to present problem & get their cooperation. Seminar on cooperatives. Organization of a Fishermen's Cooperatives. Communally owned pumpboat & nets (sharing of income).

SAMPLE GUIDELINES/SUGGESTIONS FOR GROUP DISCUSSIONS DURING SATURDAY WORKSHOPS

by D. F. Hernandez

ANTICIPATED PROBLEMS UNDER PHILIPPINE CONDITIONS	TECHNOLOGICAL/INSTITUTIONAL SOLUTIONS
(a) Crowded living conditions	- Architectural/urban design innovations -- specify - Management techniques adapted to very low per capita space/resource availabilities --specify - Behavioural technologies for conflict resolution --specify - Research/study/application of Filipino values most viable during space/resource scarcities --specify - More efficient but ecologically stable agricultural/aquacultural food production techniques -- specify
(b) Decreasing availability of lowland areas for agricultural/residential purposes; decreasing forest cover of watersheds; increasing water/wood supply requirements	- Ecological/agricultural upland management technologies -- specify - New Government regulations/programs for watershed areas "with more teeth" --specify - New research programme emphasis
(c) Greater energy vulnerability of urban areas; rural-to-urban migration pressure; predominance of western-inclined urban culture over rural/indigenous cultures; food production, a strategic national effort in the 21st century in rural/hinterland areas	- Rural-oriented technologies to accelerate productivity in rural areas faster than in urban areas --specify - Decentralized, dispersed, and cheap energy sources for rural households --specify - Institutions/programmes to reverse cultural biases militating against appreciation of some rural/indigenous cultural elements in the Philippine scene -- specify - Policies to deploy capital away from metropols and mobilize capital from within rural peripheries --specify
(d) Higher energy costs; higher transportation costs	- Policies/programmes for changing consumer life-styles/habits from energy-wasting products/services over to energy-saving ones --specify - Cheap and appropriate alternative energy sources -- specify - Energy management skills/technologies --specify - More efficient modes of transportation --specify - Energy-conserving architectural/urban designs --specify.

PILOT PROJECT FOR BIOLOGY IN THE ARAB STATES: ALECSO CURRICULUM CONTENT

by M. Saber-Selim

(slightly shortened and re-arranged by the editors)

A. 10th GRADE

Living organisms

- Historical background of Biology and its branches.
- Efforts of biologists with special reference to the role of Arab scientists.
- Methods of scientific investigation in biology through the study of some biological issues (examples from which to select: Malaria, Bilharziasis, Desert reclamation, Nutrition and some malnutrition diseases, Cotton leaf caterpillar, Soil erosion).
- Study of two ecosystems: one aquatic and the other terrestrial to deal with: Selection of site. Tools and instruments to be used in the study. Methods of study. Survey of plants and animals in the environment from the qualitative and quantitative aspects. Distribution of individuals of the same species in relation to each other and to other species. Seasonal distribution of organisms.
- Study of one example each of terrestrial aquatic animals/plants from the following points: External features, structure, mode of life, comparative study to point out the concept of adaptation.
- Environmental factors: Climatic factors (light - temperature - humidity - wind). Soil factors (Physical: texture - structure - porosity. Chemical: acidity and alkalinity - mineral substances). Effects of climatic and soil factors on organisms. Biotic factors (Effect of plants on climate: light - temperature - humidity - wind. Effect of plants on soil, mechanically and chemically. Indirect effect of animals on climate. Effect of animals on soil, mechanically and chemically).
- Relations among plants, relations between plants and animals, relations among animals (with special reference to food chains).
- The web of life: individual - group - population - community - society, with indication of behavioural adaptation of organisms according to the requirements of their environment.
- Natural biological balance and dynamics of balance; selection and succession.
- Effect of Man on environment (pollution - destruction - construction - protection).
- The world as an ecosystem - differentiation in the same environment.
- Why and how are organisms classified? - The major classes of organisms with special reference to those of economic or hygienic importance.

B. 11th GRADE

Inside the organism

Note: One of the main objectives is the emphasis on the concept of cell and its contents and functions (except reproduction).

- Components of the organism and their unity in the living world.

- Inorganic and organic components.
- The unity of organisms physically and chemically (the physical and the chemical properties of living substance).

1. Nutrition (The concept of nutrition. Nutrition in green plants. Absorption of water and minerals. Gas exchanges. Anatomy of the green leaf. Mechanism of photosynthesis. Photosynthesis and chemosynthesis in some micro-organisms. Nutrition in non-green plants).
2. Digestion (The concept of digestion. Digestion in uni-cellular organisms. Digestion of organic substances. Digestion in animals: intracellular digestion as in sponges, extracellular digestion as in humans. Digestion in plants: digestion in insectivorous plants. The human digestive system; digestive processes with indications to the hygiene of the digestive system).
3. Transport (The concept of transport. Exchange in uni-cellular organisms. Transport in plants: systems of transport, transport processes, transport in xylem vessees, mechanism of transpiration, role of phloem in transport, theories of transport of prepared food from leaf to other parts of the plants. Systems of transport and transport processes in animals: sponges, hydra, earthworm, locust, vertebrates. The human circulatory system and its physiology and hygiene).
4. Respiration (The concept of respiration. Energy and the need of living organisms for energy. A historical review of respiration - respiration and combustion. Sources of energy. Uses of energy. Energy stored in foods. Measurement of energy consumption. Respiration in plants. Respiration in animals: sponges, hydra, earthworm, insects, aquatic and terrestrial vertebrates. An anatomical and physiological study of the human respiratory system and its hygiene. Cellular respiration. Relations between respiration and photosynthesis. Fermentation; comparison between respiration and fermentation).
5. Excretion (The concept of excretion. Excretion in unicellular organisms. Excretion in plants: excretion of water, minerals and gases. Excretion in animals: sponges, hydra, earthworm, insects, aquatic and terrestrial vertebrates. An anatomical and physiological study of the human excretion system and its hygiene).
6. Support (The concept of support. Support in unicellular organisms as in Radiolaria and Foraminifera. Support in plants: herbs and trees. Support in animals: sponges, corals, echinoderms, molluscs, arthropods, vertebrates. An anatomical and physiological study of the human skeleton).
7. Movement (The concept of movement. Movement in unicellular organisms. Movement in plants: tactile movement, sleep movement, tropism. Movement in animals: hydra, earthworm, arthropods, vertebrates. A study of human muscular tissues and their hygiene).

8. Sensation (The concept of sensation. Sensation in unicellular organisms. Sensation in plants. Sensation in animals: worms, insects, vertebrates. An anatomical and physiological study of the human nervous system and its hygiene).
9. Hormonal Coordination (The concept of hormonal coordination. Hormonal coordination in plants: auxins, their types, their effects on seed germination and growth, floral hormones, abscission of leaves. Hormonal coordination in animals: study of the ductless endocrine glands in man, the hormones they secrete and the function of each hormone. Neurohormonal coordination).

C. 12th GRADE

Perpetuation of life

1. The Cell (Historical background of the discovery of the cell. The cell theory. Structure of the cell: cytoplasm and its components, nucleus. Chemical composition of nucleic acids. Some features of cell biology. Movement inside cells and between cells. Cellular metabolism: anabolism and catabolism. Nucleus and cell division. Tissues).
2. Reproduction and development (Concept of reproduction. Reproduction in viruses. Asexual reproduction: in unicellular organisms like Bacteria, Fungi, Algae, Amoebae; in higher plants, e.g. vegetative reproduction and its importance; in Metazoa like hydra, earthworm, insects, incl. multiplication of embryos in Metazoa. Sexual reproduction: Meiosis and production of gametes; sexual reproduction in unicellular organisms;in higher plants, e.g. flower, pollination, fertilization, formation of fruits, seeds and their dispersal; in Metazoa like Hydra, earthworm, insects; in mammals, e.g. human sexual glands and structure of sex cells, fertilization, role of hormones in reproduction, the menstrual cycle, pregnancy, lactation. Other types of reproduction, e.g. parthenogenesis in insects, in larvae).
3. Differentiation and Specialization of cells (An introduction to show the necessity of cell specialization in organisms in relation to their complexity, with some plant and animal examples. Study of apical meristems in plants and the differentiation of adult tissues from meristematic cells. Growth in plants. Stages of growth in frog embryo (ladople) and the differentiation of animal cells and tissues. Sexual hygiene).
4. Heredity (Concept of heredity. Heredity land environment. Mendel's experiments and Mendel's laws, with some examples. Probability in heredity. Interpretation of Mendel's laws by the chromosome theory. Non-Mendelian characters; crossing over and chiasma. Incomplete dominance and gene interaction. Hereditary characters and their relation to sex, e.g. sex chromosomes and their role in sex determination. Mode of

inheritance of some characters and diseases in Man, e.g. eye colour, height, skin colour, susceptibility for disease, blood groups. Evolution of the gene concept since its discovery up to now. Chemical basis of heredity, i.e. to show that DNA is the genetic substance. Relationship between genes and enzymes. Gene and chromosome mutations. Genetics and improving production in animals and plants. A general view of heredity).

5. Continuity of Life (Introduction including biological principles like preservation of internal=physiological and external= ecological equilibrium, unity and continuation of life, diversity and similarity from an evolutionary point of view. Relation between diversity and adaptation, e.g. diversity and adaptation, natural selection, succession of life. Variety of life in the past, e.g. evidence from fossils, review of life forms in the past during different geological periods, evolution of new types from a common ancestor; according to evidence from comparative anatomy, embryology, physiology, comparative biochemistry and heredity. Mechanism of evolution, incl. Lamarckism, Darwinism, Mutationism, Neo-Darwinism, origin of life, improving future life on the basis of biological concepts).

INDEX